AF560234

THE PLOUGH AND THE PEN

THE PLOUGH AND THE PEN

Peasantry, Agriculture and the Literati in Colonial Bengal

BIPASHA RAHA

MANOHAR
2012

First published, 2012

ISBN 978-81-7304-941-5

Published by

Ajay Kumar Jain *for*
Manohar Publishers & Distributors
4753/23 Ansari Road, Daryaganj
New Delhi 110 002

Typeset at

Digigrafics
New Delhi 110 049

Printed at

Salasar Imaging Systems
Delhi 110 035

for

Swapan

Contents

Acknowledgements

This book took a very long time to see the light of day. I have incurred several debts in the process. Over the years I have benefited from the help and advice rendered by friends and colleagues and the ungrudging service of librarians who have helped me locate my sources.

I would like to take this opportunity to thank my teacher Prof. Binay Bhushan Chaudhuri for always being there to answer my endless queries. He prompted my interest in agrarian history since I was a postgraduate student. I continue to learn from him. Sir and Triptidi will be happy to see the book. Deepak Kumar has been my source of encouragement. Without his prodding, this work would probably have remained unpublished. I have lost count of the number of times I have told him the manuscript is almost ready though never quite. I cannot adequately thank him. He has given this manuscript careful reading a number of times. He also brought to my notice the Voelcker Report. Peter Robb took time off his busy schedule to read this manuscript at SOAS and offer new insights.

I am especially indebted to the librarians, British Library and the School of Oriental and African Studies Library in London, the National Library, the Nehru Memorial Library, the Indian National Archives, the West Bengal State Archives, Bangiya Sahitya Parishad, Ramakrishna Mission Institute of Culture and Library, Centre for Studies in Social Sciences, Visva-Bharati Central Library and the Rabindra Bhavana Archives in Santiniketan, India. The latter is the greatest repository of Tagore's works. My research for this book was helped by grants from the Charles Wallace Trust and the Indian Council of Historical Research. I am grateful to Richard Alford for the interest he took in my work.

There are many friends and colleagues who have helped me in my research. Suchibrata Sen and Shouvik Mukhopadhyay have always been generous with books and advice. I have benefited from endless discussions with them. I would also like to thank Sanjukta Das Gupta and Subhayu Chattopadhyay. My students Arnab Sarkar and Madhumita Mondal helped me in locating books. I would also like to mention my students in the postgraduate classes at Visva-Bharati University. With those who specialized in agrarian history, a course I have been teaching for years, I have discussed many of my ideas that have found expression in this book. These interactions have helped to keep alive my interest in the subject.

My debt to my family is too great to be enumerated. My mother has been my constant source of support. So has been my mother-in-law who has never complained about the long hours I devoted to my research. It will be my eternal regret that my father and father-in-law did not live to see the publication of this work. My sister Manisha was always encouraging. My sons Arka and Aritra have frequently reminded me that there is more to life than agriculture and peasantry. Rimi, Roni, Subho and Shayan have been a constant source of joy. My husband Swapan has been by my side, as usual, throughout the writing of my book, always encouraging me to get it done. Eternally patient, he has helped with his suggestions, read drafts and done much of the typing. Now I can take the train to Santiniketan without having to think about agrarian history.

BIPASHA RAHA

Abbreviations

B.S.	:	Bangla San (Bengali Year)
BPP	:	*Bengal Past and Present*
BIA	:	British Indian Association
GOI	:	Government of India
IESHR	:	*Indian Economic and Social History Review*
IHR	:	*Indian Historical Review*
Misc.	:	Miscellaneous
n.d.	:	Not dated
n.m.	:	Not mentioned
Proc.	:	Proceedings
Rev.	:	Revenue
RNP	:	Confidential Report on Native Papers (Bengal)
RR	:	*Rabindra Rachanavali*

CHAPTER 1

Colonial Rule: The Changing Rural Landscape

From the Mekong to the Gangetic valley, the nineteenth and early twentieth centuries witnessed protracted struggle of the peasantry against generations of exploitation. In societies, groaning under the throes of colonial domination it acquired, in particular, a new dimension. There was increasing resistance by peasants against exploitation by the colonial state and the dominant landholding and moneylending and merchant classes. Distinct signs of a growing consciousness informed the activities of subordinate sectors of the agrarian society. Daily resistance to oppression and occasional outbursts of protests, at times with great effect, became the usual practice. They threatened the existing structures of domination. In the process, they set in motion a series of developments that impacted upon and completely recast the emerging Bengali literati's perception of agrarian issues and rural life.

The nineteenth century saw the emergence of the modern intelligentsia in Bengal. It was here that British rule was first established and modern Western education introduced. They were a product of this new education and the land settlement introduced by the new rulers. The literati found it increasingly difficult to remain oblivious to the multifarious problems besetting rural agrarian society. With the dawn of political consciousness and the beginning of anti-colonial movements, interest in the economic problems of the country accompanied the growth of political aspirations. The hitherto insurmountable distance between the downtrodden peasantry and the potentially powerful, predominantly Western educated and increasingly numerous middle-class intelligentsias seemed to lessen considerably, as a section of the latter strived to emerge as the natural leaders of society by championing the cause of the former and articulating their grievances. In this way some hoped to create a mass-base necessary for the fulfilment of some of their political aspirations.

Agrarian thinking at times, became a part of organized politics. In the nineteenth century, significant changes took place in Bengal's agrarian economy and the rural countryside. The contemporary literati were not always in a position to comprehend the actual implications of these changes. Some of these changes were often not immediately visible to them.

MAJOR AGRARIAN CHANGES AS A CONTEXT OF THE CHANGING PERCEPTION

Major agrarian changes were the outcome of the overall British economic policy. These changes started to occur in the last quarter of the preceding century. Consequently, the agrarian economy, which emerged in Bengal in the nineteenth century, was different in many respects from the one, which the British had inherited. All these changes had partly to do with the determined policy of the colonial government to maximize their revenue. Their aim was to extract as large a part of the surplus in the form of land revenue, as was possible from the empire they created in India. With this aim in mind, they attempted a reordering of rural society. Recent researchers have added immensely to our knowledge of the changes affecting rural Bengal throughout the nineteenth and the first quarter of the twentieth century and the reasons for these changes. It is proposed to analyse the broad changes affecting Bengal's agrarian economy as the contemporary Bengali literati viewed them. Recent studies have questioned the validity of the some of their ideas.

The Permanent Settlement fixed the land revenue in Bengal permanently. Land was assessed at a very high rate. This, however, did not mean a complete freezing of land revenue, and the Company could secure an increase from time to time.[1] The largest part of the increase came from the resumption of 'rent free' lands. The Settlement recognized zamindars as having an unqualified, proprietary right in the soil alone. It gave them 'all property rights which could not be proved to be an encroachment on the prescriptive or customary rights of the tenants'.[2] The 'prescriptive' and 'customary' rights of the tenants were not defined. It simply defined the relation between the zamindar and the government, and left the subject of the relation between the zamindar and the ryot open for later legislation. This system on the one hand created a class of landowners dependent on the political authorities for its existence and on the other, made suitable arrangements for easy collection of revenue. The zamindar's position thus underwent a change. They were now elevated to the position of landowners. A class of free landholding peasantry became their tenants, who could be easily evicted by them on their failure to pay their rent on time or at the enhanced rate.

There was considerable dislocation in the rural society at least in the initial years. The zamindars had to pay 9/10 of the rent collected from the ryots to the government as revenue. This was a rather steep rate at that time. In the early stages, many zamindars lost their zamindaries because of their inability to pay the revenue in time. There was considerable confusion in revenue management.[3] The zamindars refused to issue *pattah*s. They were afraid that these could be interpreted as a limitation on their proprietary rights. The tenants also failed to press their customary rights in courts. The early years was a period of favourable land-man ratio.[4] So when the ryots refused to pay their rent or abandoned their holdings and moved elsewhere, many of the estates were sold for arrears of revenue. Many of the old zamindari families were ruined. Moneyed men from urban areas acquired new zamindari status. This distress sale of lands resulted in the formation of a land market. Many of the new zamindars, men who had mainly made their profits through trade with the Company, were inexperienced in zamindari work. They preferred to leave the work of collection of revenue to the *gomastha*s and *amlah*s and live in the town. The old and new zamindars also created different types of rights in land to facilitate collection of revenue. Consequently, different intermediary tenures were created. None of these intermediary tenure-holders was interested in the improvement of land. All they were interested in was land revenue.

In order to enable the zamindars to collect their dues the *Haftam* and the *Panjam* regulations were passed in 1799 and 1812 respectively.[5] These empowered the zamindars with coercive powers to extract rent from the tenantry. The legal sanction given by another regulation in 1819 to the creation of *patni* tenures gave the zamindars further powers. Gradually, landed proprietors became more and more a closed community,[6] admitting new members much less frequently than before. This was due largely to the reduction in the auction of lands.

The condition of the ryots continued to deteriorate. Due to the gradual destruction of the traditional cottage industry, the increasing population became more and more dependent on agriculture. Not much improvement in land was undertaken. The influence of the moneylenders was on the rise. The ordinary ryots depended on them for meeting the rental demand and for buying plough animals and seeds. They often mortgaged their lands to borrow and could never pay off their mortgage. As a result, a part of their land used to be transferred into the hands of the moneylenders.

In the course of the nineteenth century, the government made two significant attempts to legally amend some of the provisions of the Permanent Settlement of 1793. One was the Rent Act or Act X of 1859 and the other was the Bengal Tenancy Act of 1885. The failure of zamindars to emerge as a class of entrepreneurs, the need to promote peasant enterprise in agriculture by restricting the powers of the zamindars to rent enhancement, increasing rural tension and recurring famines prompted such actions.

The Rent Act intended to remove some of the abuses of the existing laws, particularly of the laws relating to distraint of crops and properties of the peasants for the realization of arrears of rent. This Act lay down that the zamindars could not any longer compel attendance of peasants at their courts for the adjustment of rent or for any other purpose. Distraint was illegal for arrears of more than one years' standing or when the peasants could provide security for the payment of arrears within a reasonable limit of time. The most important provisions were those relating to the rent question. Ryots were classified into three groups: ryots holding at fixed rates, occupancy ryots and non-occupancy ryots. Ryots holding land at a rate unchanged for twenty years before the date of rent—suits brought against them by zamindars belonged to the first group. Zamindars could not enhance their rent. Any ryot who had cultivated or held lands for a period of twelve years was an occupancy ryot, as long as he paid rent. Zamindars could enhance their rent only where cultivation had increased, or where the rent rate paid by them was lower than what was called the *pargana* rate (the prevailing rate for lands of a particular quality) or where the value of the produce had increased. These legal restraints would not apply where occupancy ryots agreed by a written contract to pay an enhanced rent. The law did not protect ryots without the right of occupancy. The Act also permitted the peasants, in dispute with the zamindars over the rate and amount of rent, to deposit their due rent at the existing rate with the collector.

The Rent Act did not serve the interests of either the landlords or the tenants. Under the Act, the new occupancy ryot was required to furnish the almost impossible proof that he had held every particular field of his holding for twelve years.[7] The zamindars could also easily evade the law by shifting his tenant from one field to another in order to prevent the accrual of occupancy right. The law also allowed the ryot to contract himself out of the occupancy rights. The zamindars resented it especially as it curtailed their power of rent enhancement. In the period following the enactment of Act X, the landlords carried on a prolonged crusade against it. They resorted to illegal means in order to enhance rents and tried to deny occupancy status to those entitled to it and were very often successful in their efforts. In view of the increase in the number of taxes and cesses imposed on them like the Public Works cess, road cess, *chowkidari* tax, etc., increase in the rental income was necessary particularly because landed income was their sole means of survival. Two other reasons for the increase in rent rates were peasant resistance to *abwabs* and apprehension that the new rent law would fix the rent rates permanently.[8] Act X neither gave the occupancy ryots any protection from incessant enhancement nor determined any rules for fixing an equitable rate of rent.[9] The law also did not define the occupancy ryots' right to improvement. Nor did it specifically determine his right in them in the event of his eviction.[10]

The Rule of Proportion enunciated in a judgement of 1864, by which the size of the enhanced rent was determined, was often practically unworkable after the transference of rent suits from revenue officers to the civil courts in 1869. The enhanced rent the Rule laid down bore to the previous rent the same proportion that the increased gross value of the produce bore to the previous gross value. The onus of establishing their case by full legal proof lay entirely on the zamindars. As long as the revenue officers tried such cases, zamindars won many of the suits despite their failure to provide all the necessary evidence because the revenue officers seemed to have entered upon the cases with the knowledge, that as a rule, prices of produce had risen. With the transference of the rent cases to the civil courts, the position of the zamindars became more difficult. The courts, unlike revenue officers did not assume the fact of the price rise but relied exclusively on the ordinary rules of evidence.[11] Zamindars thus found it extremely difficult to prove, when the rent was previously fixed and what was at the time the gross value of the produce. Thus failing to enhance rent by legal means, the zamindars sought to make up for the loss by illegal exactions, enforced in several cases by illegal means.[12] The truth of this was established by an official enquiry during Campbell's administration (1871-4). However, he was reluctant to reconstruct Act X on new lines.[13]

His successor, Richard Temple in view of the increasing tension between the zamindars and the ryots felt it imperative to introduce a law facilitating speedy settlement of rent disputes. On 10 July 1876, the Agrarian Disputes Act V was passed. This transferred rent enhancement suits from the civil courts to the hands of the collectors. Temple felt that, the collectors were better equipped to deal with rent suits in view of their greater familiarity with the intricate questions relating to rent.

Tenancy Act VIII of 1885 was the next important milestone in the history of rent legislation. This Act made the acquisition of occupancy rights a much easier process. A peasant could qualify by cultivating for twelve years any plot of land in a given village. It was no longer necessary to continue cultivating the same plot of land. This would secure the peasants against the practice adopted by the zamindars of changing plots of lands every five or six years to prevent the former from acquiring occupancy right. Unlike the Act of 1859, the new Act did not allow peasants to contract themselves out of such a right. It deprived zamindars of their power of ejecting peasants for arrears of rent alone. The sale of an occupancy ryots' holding in execution of a decree for the rent was, however, permissible. As for the question of rent, the Act did not reduce the present quantum anywhere. There was no change either in the three main grounds of enhancement of the old Act. The government would henceforth be responsible for the preparation of the price list necessary for deciding enhancement suits brought by zamindars on ground of an increase in the value of produce. Moreover, rent could be enhanced only once in a period of fifteen years. The quantum of enhancement would not be more than 12.5 per cent of the existing rent. There was not much change in the incidence of occupancy rights. The Act did not make provisions for free alienation of holdings. However, where such alienations had been an established custom, the civil courts would not interfere with them. The Act also gave the government the authority to prepare a Record of Rights based on survey even in the zamindari estates. All these provisions, however, applied only to the class of occupancy ryots. The non-occupancy ryots and the under tenants of occupancy ryots were left entirely unprotected.

An elaborate system of rights to land including the property right in revenue collection granted by the colonial state accounted for the consolidation of economic and social power within rural agrarian society. Recent researches have identified major types of social organizations of production in the nineteenth century—the demesne-labour complex, the peasant small holding complex, the rich farmer-sharecropper arrangement, the plantation and the tribal communitarian form.[14] The settled ryot paying his rent in cash constituted the backbone of the agricultural population. In most east Bengal districts, peasant smallholding was the predominant form of social organization of production.[15] However, demographic pressure between 1890 and 1920 saw the emergence of a land-poor peasantry dependent on sharecropping and wage labour. In much of west and central Bengal rent collecting landlords, drawn mainly from upper Hindu castes, supervised farming on substantial chunks of land which they held as their personal demesnes or *khas khamar*. Small holdings of peasants drawn largely from middle agricultural castes like Mahishyas, Sadgop and Aguri were also there. Low caste Bagdi and Bauri as well as Santal tribal labour worked the *khamar* lands. The increasing grain prices from the mid-1850s onwards contributed greatly to increasing differentiation within the peasantry. The rentiers within the small holding demesne complex were not only substantial landholders but creditors as well. So were the richer peasants, who too were involved in moneylending and grain dealing.[16] The plantation system was common in north Bengal districts, viz., Dinajpur. Under the impact of colonialism, significant changes also took place at the point of production.[17]

From 1770 to about the middle of the nineteenth century, all parts of Bengal was subject to population increase which was more marked in some regions than other.[18] During the period between 1860 and 1890 east Bengal witnessed a rise in population and rapid expansion of cultivation and total output. The Muslim and Namasudra peasants were mainly responsible for the extension of cultivation.[19] Extensive lands were brought under cultivation.[20] There was double cropping in many districts of Bengal.[21] However, persistence of traditional agrarian practices affected agricultural productivity.[22] The overall picture during this period shows stagnation in agricultural production.[23] From the 1920s overall better control of disease brought about a sharp fall in the death rate in west Bengal. In east Bengal, the population pressure continued to increase on land while output did not show any noticeable increase and at times often stagnated.

In this period, as recent researches emphasize, agriculture was adversely affected in west Bengal both by the 'exhaustion of the land of the moribund delta and by the mortality as well as morbidity of labour' because of malaria infection.[24] Gross output in west Bengal declined between 1860 and 1920 in a period of stagnation of population, though there were some pockets of growth, viz., in the 24-Parganas, Khulna and parts of Midnapur.[25] Demographic trend in eastern parts of north Bengal was very similar to that in eastern Bengal. However, it was the immigration of mostly Santal, Oraon and Munda tribal labour from Bihar, which boosted population level in the western parts of north Bengal between 1860 and 1920.[26] Immigrant labour played a crucial role in the clearing of jungles in this part. The tea gardens in Jalpaiguri and Darjeeling also recruited immigrant labour on a large scale. In east Bengal, the birth rate continued to be marginally higher than in west Bengal and the death rate considerably lower. Increasing population in east Bengal was mostly a natural increase based on large surplus of births over deaths.[27] This was not so in north Bengal where immigrant labour from tribal habitats in Bihar[28] set to work as sharecroppers and plantation labourers.

A significant development of this period is increasing commercialization. Exposure to world economic forces and the compulsions of the colonial state influenced the working of the agrarian economy. Increasing commercialization and monetization in the mid-nineteenth century has been variously explained. An early view had credited increasing commercialization to the colonial state's inflexible demand for land revenue payable in cash and agricultural production for the market.[29] A more recent view[30] while not denying the revenue-money-commerce nexus cautions against acceptance of any notion of an inexorable sequence set in train by the cash-hunger of the colonial revenue establishment by pointing to the levels of monetization and commercialization reached in the immediate pre-colonial era. Yet, the colonial state was a key factor in bringing about major shifts in the scale and character of commodity production. However, some of the impetus in this process came from the centre of an emerging capitalist world economy based in Europe.[31]

One illuminating research has highlighted some general circumstances affecting the growth of commercial agriculture in Bengal in the later half of the nineteenth century.[32] Apart from the particular circumstances contributing to the growth of the cultivation of different cash crop, several developments in the world in the later half stimulated India's foreign and internal trade in general trade largely consisting of raw materials and agricultural produce.[33] The liberalization of tariff policy by the Government of India, particularly after 1867, by abolishing or reducing export duties on many commodities, and the gradual fall in ocean freight also contributed to the expansion of India's foreign trade. The internal trade and commerce was stimulated by a gradual development of communications, particularly the railways which helped carry agricultural produce from village to distant markets.[34] In Bengal in the nineteenth century European capital was invested in the plantations of Darjeeling and Jalpaiguri.[35] Opium cultivation was a state monopoly and mulberry production was almost entirely state financed.

As regards the nature of commercialization in production of indigo in west Bengal in the first half of the nineteenth century, it was stimulated by the increased demand in Europe in late eighteenth and early nineteenth century and its importance as a means of remittance. Of the two main forms of indigo cultivation, on *raiyati* or peasant lands and *nij* or demesne land, where the planter had bought into a superior tenurial right, the former was much more prevalent.[36] Before 1825 indigo cultivation was not a forced one.[37] After 1830 apart from outright coercion,[38] the only reason for wanting to cultivate indigo was the money advance but they could not repay it.[39] No indigo peasant could hope to redeem the unpaid balances which mounted over the generations, thereby attaching unpaid labour to indigo cultivation.[40] Once, the planters had become wary of making new advances to the peasants, the only reason for wanting to cultivate indigo disappeared.

The outbreak of the Crimean War (1854-6) cut-off the supplies of flax and hemp from Russia and created a new demand for Bengal jute.[41] Demand for rice in the European and Chinese markets, as well as diversion of some rice lands to produce jute and oil-seeds, set an inflationary trend in the rice price beginning in 1855. The price of rice increased steadily in Bengal between 1855 and 1860.[42] At a time when there was a general increase in prices, food grains in particular, the forced cultivation of a wholly unremunerative crop like indigo came to be increasingly resented by the peasants. They were encouraged by the indigenous moneylending landlords who now saw better prospects in the rice sector. The planters were defeated in Bengal. But moneylending landlords and traders were to emerge as major claimants of the peasant's surplus in a phase of subsistence commercialization from the later nineteenth century.[43] However, indigo cultivation did not entirely disappear after the revolt in Bengal proper.[44] Bengal continued to produce nearly one-third of what Bihar produced.[45] A new feature of the cultivation in Bengal was the increasing indigenous enterprise.[46]

In the second half of the nineteenth century, population pressure, subdivision of holdings, diminishing returns from land in the context of persistence of traditional agrarian techniques and dwindling incomes forced poor peasants to adopt the cultivation of high value and labour-intensive cash crops in an attempt to increase income. Jute was a new crop. Its emergence as an important cash crop dated only from the mid-1850s. The demand for jute in the world market increased particularly since the Crimean War.[47] In spite of a brief downward trend in the 1870s because of famines, the rapid expansion of international grain trade during the final quarter of the nineteenth century, provided the stimulus for the growth of the acreage under raw jute as well as the establishment of a jute manufacturing industry in east India. There was also a home demand for container bags. Bengal jute entered its most vigorous and sustained period of growth which lasted until the outbreak of the First World War.[48] The total value of Bengal's jute, however, maintained an upward trend until the world depression of 1930. The market of raw jute tended to widen as a result of its increasing consumption in the new jute mills in Calcutta, its neighbourhood and Serajgung for the manufacture of gunny bags, and of its increasing exports. During the last quarter of the nineteenth century we also find the increasing popularity of yet another commercial crop—tea in north Bengal. Tea plantations, owned by Europeans, were run mainly with the help of tribal labour.

Cultivating a cash crop which promised a higher gross income, seemed to the small holders, a better option in the early twentieth century, than growing insufficient quarters of rice.[49] However, it must be emphasized that in a scenario of predominantly small holding cultivation, the overwhelming majority of primary producers cultivated cash crops under compulsion.[50] From the early nineteenth century mostly small holding Bengal peasants engaged in expanded commodity production for a capitalist world market. That did not necessarily entail capitalist transformation of the complex relations of property and production in agriculture.

Anybody with money to buy intermediate land rights could buy them and establish a direct claim on agricultural surplus.[51] An inevitable part of this process was the growing extent of rural indebtedness. The peasants had to take increasing recourse to borrowing, in an attempt to maintain the bare minimum level of consumption. The resulting indebtedness, in which the vast majority of small peasants were caught, set in motion a different process of commercialization of agriculture. This was what some have called the 'second phase' in the commercialization of Bengal's agriculture.[52]

Commercialization forcibly involved the peasants into market relations through the mechanism of debt. Credit played a vital role in material production and in the social reproduction of peasant labour. With the rapid expansion of commodity production in the rural countryside there was increasing demand for credit. Debt obviously followed. The operation of credit as the principal mode of exploitation and the size of its appropriation varied according to the nature of the agrarian social structure.

Sugato Bose has distinguished between the *dadni* and *lagni* forms of credit representing merchant and usury capital in the countryside with regard to peasant small holders of east Bengal.[53] Both took shape from the later nineteenth century within the framework of a widening market and rising prices for the peasants' product. Merchant capital, which originated in the areas of high finance in the metropolis, was exchanged and forwarded by the purchasing firms through a network of commission agents to the primary producers. The peasant usually received his *dadan* or advance from a small trader-moneylender directly dependent on the flow of funds from above. The primary interest of the entire hierarchy of lenders was to secure the crop at a low village-level price. In addition to trader-*mahajans*, some rentier landlords who were unable to enhance rents or expand cultivation emerged as usurers in the jute-growing areas. Merchant and usury capital were distinct, but both exploited the working peasantry. *Talukdar-mahajans* were able to switch from rental to usurious income because of the expansion of the product market. Trader-*mahajans* were helped where peasants were not tied by *dadan* in procuring the produce cheaply by demand for interest or *lagni* at harvest time. In some regions of western Bengal the landlord and some richer peasants constituted the chief group of creditors. Lending within the demesne sector was mainly in grain. Small peasants could not survive without borrowing and paid a heavy price in interest payments. They were caught in a cycle of impoverishment in the process. Moneylending landlords were involved in the rice market.[54] In the jute regions in north Bengal, some trader-*mahajans* engaged in the *dadan* business. But grain loans constituted the bulk of credit, which gradually came to exercise effective control over land and the product market.

Bose observes that evidence from the early twentieth century strongly suggests that appropriation through debt interest was larger than rent and that the differential widened during the fifty-year span from 1880 to 1930. Indebtedness, on the other hand, was on the rise.[55] Sharecroppers and labourers paid for credit with their labour and a good part of the product of their labour. Credit thus played a central role in the ability of dominant social groups to appropriate surplus in a variety of ways. For most peasants mounting debt meant they were hopelessly ruined and 'in a state of manifest insolvency'.[56]

CONDITION OF THE PEASANTRY

An important development of the period was the increasing differentiation among the peasantry. A class of rich peasants emerged at the top of the ladder. Apart from their income from the land, they accumulated a considerable sum from grain dealing and moneylending. The relative richness was not a product of any superior technique of cultivation. Their origin can be attributed to some greater differential natural advantages, such as better soil, better facilities for irrigation, nearness to the markets, superior agricultural skills, enterprise, readiness to take up reclamation of wasteland and to cultivate richer crops.[57] A considerable class of sub-ryots also emerged. They leased the land from the richer peasantry who often dissociated themselves from agriculture in lieu of rent. There emerged also a class of poor peasantry. They had lost their rights of occupation over the lands which they had cultivated for the richer peasantry. However, this very seldom led to absolute eviction. The dispossessed peasants in most cases continued to cultivate the same plot of land, with inferior rights and higher effective rents. At the bottom of the ladder were the agricultural labourers who had no land and no implements of agriculture. Zamindars as well as substantial peasants employed them.

The main reason for land transfer was indebtedness. There was considerable rural impoverishment. Peasants borrowed in order to pay the rents as well as to buy grain during the lean seasons. The debts kept on accumulating. The fact that they had to sell their grain during the harvest season to pay their dues meant lower cash returns. They did not have sufficient resources to take advantage of the lean season. The pressure of population on land because of the increase of population as well as the lack of other avenues of sustenance weakened the bargaining capacity of the tenants. The low level of technique stood in the way of an increase in output. There was increase in output but that was mainly due to the extension of cultivation after the permanent settlement. However, the increase in population soon outstripped the extension of cultivation. Pressure of population soon resulted in the village common being put to the plough. Agriculture remained stagnant. Lack of adequate capital in the hands of the peasants naturally meant the absence of productive investments on land. This was particularly noticeable in the case of irrigation. The peasants lacked the adequate means to undertake measures for better irrigation. This severely handicapped agriculture in the region of an uncertain rainfall. The zamindars who were customarily responsible for irrigation and embankments neglected their duty. Besides, they also refused to allow the peasants to undertake any such measures as they could for fear of the consolidation of the occupancy rights by the peasants.

The condition of agriculture as well as of tenants on estates divided into coparcenary shares was even worse. There were constant feuds among the co-sharers of joint estates. The tenants on these lands suffered more as they were subject to extortion by all the co-sharers. The breakup of zamindari property was detrimental to the interests of agriculture. The co-sharers intent on scoring over their rivals neglected productive investments on land. The laws of succession among the Hindus and Muslims made it impossible for estates to remain undivided. The contentions and differences among the ryots were submitted to the arbitration of the joint holders, who derived some benefits from them and were, therefore, ready to foment these differences. After bringing the most influential of their tenantry under their influence by remission of their rents, or making them participators in the process of extortion, they extorted *abwab*s from petty ryots. On the occurrence of disputes among them, the co-sharers espoused the cause of different parties and put others to extreme trouble and inconvenience. The ryots in their turn, stuck to the cause of different co-sharers and thus exposed themselves to the ire of the opposing parties. The tenants were made to pay their rents more than once to the co-sharers, those who were less powerful than others, especially females and minors, suffered most. The stronger parties encouraged the ryots even to set aside their claims and were in a manner dispossessed of their estates.[58] Litigation was frequent in coparcenary estates, but often failed to secure justice to the weak. Both landlords and ryots suffered.

PEASANT RESISTANCE

A significant phenomenon was the increasing resistance by peasants against exploitation. An undercurrent of everyday resistance to inequities and periodic demonstrations of effective resistance, which dismantled the established structures of domination, influenced the course of agrarian history. The subordinate sectors of small holding and labouring society began to show distinct signs of a growing consciousness.

Between the 1810s and 1850s peasant grievances concerning rent and indigo came to the forefront in movements of agrarian resistance. So long as the demographic situation remained relatively favourable, desertion and migration were the most common strategies resorted to by peasants anxious to avoid paying enhanced rents. Migration and desertion by peasants as a form of resistance to rent/revenue enhancements and extortions has a long history. Its beginning may be traced to as early as the Mauryan period. However, the scope for such forms of resistance declined with the shrinkage of availability of cultivable lands over the centuries. There was then greater resort to revolts as forms of resistance. Refusal to fulfil indigo contracts after taking advances was the usual early form of resistance against unremunerative-forced cultivations. Changes in the broader political structure and economic context during the 1850s bought some new pressures but also opened unprecedented opportunities for effective peasant resistance.[59] The indigo-growing peasants could take advantage of the planters' loss of support among government officials and dominant classes in the rural areas as well as the urban intelligentsia. The indigo revolt spread from across the districts of Nadia, Jessore, Murshidabad, Faridpur, Malda, Pabna and Rajshahi. Expression of discontent against unremunerative indigo cultivation had been heard throughout the later 1850s, but what signalled the outbreak of the indigo revolt in 1860 were some acts of the Magistrate Eden of

Barasat which was the immediate cause. In March 1859, he upheld the peasants' right to grow whatever crops they preferred. In August 1859, he stated further that the police had a duty to protect peasants in the possession of their land even if they had broken contracts.[60] Once this was made known to the peasants, rebellion broke out. Supported by the small *talukdars* and the urban intelligentsia, the rebellion soon spread. It received a temporary setback when the government revived the Regulation of 1830. But it was withdrawn in September 1860. Though the indigo system in Bengal was fatally affected by the revolt, it also left the moneylending and grain-dealing landlords as the dominant class in rural West Bengal.

Act X of 1859 brought about a change in the nature of peasant resistance. The main issues of conflict between landlords and tenants became occupancy rights and rent during the anti-rent agrarian movements in east Bengal in the 1870s and 1880s. The source of the conflict in east Bengal was the attempt by some rentier landlords to disregard the pro-tenant provisions of the Rent Act to increase rents in an era of rising prices. In particular, these landlords sought to prevent the accrual of occupancy rights, which would have assured moderation of rents, by shifting their tenants around different holdings in the village and getting them to accept new *kabuliyats* foregoing their legal rights to occupancy status. Unable directly to serve notices of rent enhancement, they also tried to consolidate various legal *abwabs* with the juridical rent in the new *kabuliyats*. The peasants easily saw through the manipulations of the landlords and joined forces to resist them. In May 1873, an agrarian league was formed in the Yusufzahi *pargana* of Pabna district to fight against the impositions of the Banerjee, Tagore and Sanyal zamindari families of the area. Although the leading historian of the Pabna disturbances denies the official view[61] that the prosperity of the ryots was a causal factor in the tensions, according to one recent analysis, the battle was at least in part about the share-out of the higher gross income brought in by jute.[62] The central theme of this protest movement was the refusal to pay rent. The militant phase of the Pabna protests lasted from May to December 1873 and had a powerful demonstration effect on the peasants in other districts of east Bengal. Agrarian leagues on the Pabna model organized rent strikes in the mid-1870s in Bogra, Rajshahi, Dhaka, Mymensingh, Faridpur, Tipperah and Bakharganj against what has been termed as 'high landlordism' by one historian.[63] He has defined it as a stage of change in the pattern of ownership and its impact on the tenantry which developed under nineteenth-century colonization, of the growth of a desire on the part of the some landlords to hold their estates untrammelled by even conventional rules and regulations for protection of their tenantry. These agrarian protests demonstrated the power of agricultural unionism, underlined the basically unstable nature of landlord-tenant relationship and sharply pointed to the inadequacy of the existing law which governed these relationships.[64] In the anti-rent agitation we find that class interests predominated over religion.[65] There was no other major insurrection in the nineteenth century.

MAJOR TRENDS IN THE FIRST QUARTER OF THE TWENTIETH CENTURY

The permanent zamindari system, which survived throughout the nineteenth century, began to crumble in the twentieth. Population continued to rise. In early twentieth century, in east Bengal the birth rate continued to be marginally higher than in west Bengal and the death rate considerably lower.[66] Increasing population resulted in continuous reclamation of unoccupied wastelands as well as declining per capita output. Between 1920 and 1946, for instance, total agricultural output grew at a mere 0.3 per cent per annum and food crops output stagnated, while population increased at an annual rate of 0.8 per cent.[67] Even in the areas of aggregate agricultural growth, per capita output declined.

The condition of the ordinary peasants kept deteriorating. Due to the destruction of the handicrafts industry, a large section of the gradually increasing population became dependent on agriculture for their livelihood. However, there was no improvement in the techniques of agriculture. The influence of the moneylenders increased in the rural countryside, as the poor peasants depended on them to meet the expenses of buying plough animals and seeds, and pay revenue on time. They mortgaged their lands to them, which they would never recover. As a result, a portion or the entire land of the peasant became the property of the moneylender.

The total value of Bengal's jute showed an upward trend until the worldwide depression of 1930.[68] However, the jute producers did not benefit from the increase in jute acreage. The manufacturers and exporters benefited the most. They now increasingly tried to maintain an assured supply of the raw product at a low price by financing agrarian production and trade. The annual inflow of foreign finance and merchant capital became a crucial element in Bengal's commercial agriculture. Usury capital was still important in the countryside. The usurer was often a rentier landlord.[69] The interest rate on *dadan* ranged between 24 and 75 per cent and the debtor undertook to deliver the crop at 10 to 25 per cent below the market price. As one historian has pointed out, 'the cultivators were denied fair prices, not primarily because of the intervention of too many middlemen, as alleged by British civilians', but 'mainly due to combination of European buyers of jute'.[70] The jute purchasers used different techniques to deprive the peasants of a fair and equitable price.[71]

Rice continued to dominate as the chief subsistence crop with considerable commercial importance. The bulk of the rice trade was inter-district in character.[72] The Dutt Committee's detailed investigation of prices between 1890 and 1912 had found that the growth of agricultural income had been sluggish compared to the rise in the cost of living.[73]

In the early twentieth century, the peasant small holding system continued to be the predominant form of social organization of production in most east Bengal districts. However, the increasing pressure of population between 1890 and 1920 saw the emergence of a land-poor peasantry dependent on sharecropping and wage labour. The landless labourer was reportedly rare anywhere in eastern Bengal. Dependence on wage labour in early twentieth-century east Bengal was, in fact, not as rare as landlessness.[74] A considerable section engaged in short-distance migration from poorer to better-off districts to harvest jute and paddy where the timing of crop cycles permitted.

During this time indebtedness within the peasant small holding system widened because of the increased market penetration and the enhanced credit needs of the peasantry.[75] The peasants needed cash not only to tide over the lean periods, but also to carry on production. While some moneylenders dispossessed their peasant debtors for default, the usual practice was to preserve the peasant small holding system.

In much of west and central Bengal, rent-collecting landlords, including the large number drawn from the upper Hindu caste, supervised farming on substantial chunks.[76] Hired labour was in use. The rentier class and the rich peasants were involved in moneylending and grain-dealing business. Peasant small holders borrowed money to purchase food and seeds and to pay rent and labour charges.

Regular grain advances attached the labour of sharecroppers and wageworkers for the *khamar* sector. In north Bengal, it was the *jotedar* class, which was predominant. These *jotedars* were not rentiers under the Permanent Settlement, but rich farmers who after 1885 often held the *raiyati* right to their very substantial holdings. Many of them, both Muslim and Rajbansi, were deeply involved in the credit and product market.

In this period a large section of the landholding class also suffered as they were owners of small farms. The number of powerful zamindars or intermediary landholders was very small. Division of property due to inheritance law and creation of intermediary tenures led to fragmentation of landed estates, which in turn was responsible for decline in landed incomes. Zamindars with minimum incomes of Rs.10,000 got the right to elect representatives to the Bengal Legislature. At the turn of the twentieth century, their number was only one thousand. The condition of the majority of the small zamindars was deteriorating due to continuous price rise on the one hand and contraction of opportunities of increasing revenue on the other. Failure to pay their debts was often responsible for transfer of portions of their lands to the rich peasants. It was only natural that, given these circumstances, the zamindar class would try to cling on to their old prerogatives.

In the 1920s, the Bengal Legislature proposed to reform the Bengal Tenancy Act. Three issues had become crucial—the questions of giving occupancy right to sharecroppers and; under ryots and; giving occupancy ryots the right of land transfer. The zamindars opposed any attempt to limit their prerogatives. A large section of the Bengal zamindars became members of the Council with the support of the Swarajya Party. At the close of our period, the Amendment of the Tenancy Act was passed in 1928.

The occupancy holdings were declared transferable in whole or part, subject to a transfer fee amounting to 20 per cent of the sale price or five times the rent.[77] The landlord was given a right of pre-emption on payment of the sale price plus 10 per cent as compensation to the purchaser. He also retained the right to levy a fee for the sub-division of holdings in the case of part transfer, because the Act did make it incumbent on the landlords to divide the holdings in such cases. In order to prevent land from passing to the mortgage for indefinite periods, occupancy ryots were allowed to give usufructuary mortgages only for a period of 15 years. Occupancy ryots were given all rights in trees. The right to commute rent in kind into cash rent was abolished. Under-ryots were divided into three classes. Under-ryots who had already obtained rights of occupancy by custom were given the full rights of occupancy ryots, except transferability and the right to be deemed 'protected interests' against superior landlords of the ryots. The second class consisted of under-ryots who had a homestead on their land or had occupied it for 12 years continuously, or had been admitted in a document by their landlords to have a permanent and heritable right. This class could be ejected if they failed to pay their rent or if they misused the land. The third class of under-ryots could also be ejected on the additional ground that the ryots wanted the land for their own cultivation. When the Bill was introduced, it contained an explanation that for the purposes of this section, cultivation by the ryot himself did not mean cultivation by *bargadar*. However, this explanation was omitted from the Bill by the legislature. The initial rent of under-ryots was left to contract, subject to the provision that it could not exceed one-third of the estimated value of the gross produce. However, once their rent had been fixed, it could only be enhanced under a registered contract by 4 *annas* in the rupee. Persons who under the system generally known as *adhi*, *barga* or *bhag* cultivated the land of another person was not a tenant unless he had been admitted in a document by the landlord or held by a civil court to be a tenant. A person who cultivated the land of another person on condition of delivering a fixed quantity of produce was, recognized as tenant, whether he was a ryot or under-ryot as the case might be under the Act.[78]

By the Act, the sharecroppers and agricultural labourers were completely precluded from acquiring the status of ryots. Section 4(a) of the Act recorded the ineligibility of the sharecroppers to be recognized as tenants. The Act not only forbade the *bargadar* from acquiring any tenancy right but those *bargadars* who had acquired certain rights under the record of right prior to the passing of the amendment Act were also prevented from acquiring their rights. The tenants on the *barga* land were, at the same time, subjected to an exorbitant rate of rent.

THE EMERGENT INTELLIGENTSIA

The Permanent Settlement thus established a new order of landlordism. In place of the old landed aristocracy, a new class of hereditary zamindars was created. The new zamindars were responsible for further division and subdivision of landed estates.[79] *Jotedars*, *talukdars*, *grantidars*,[80] *patnidars* and other kinds of sub-tenure holders created by the zamindars increased in number. Along with them there were also the *banias* associated with the English Company, *gomastas*, brokers and small traders. All these came to constitute the middle class. They mainly belonged to the higher castes of Hindu society. They saw in Western education an opportunity to improve their prospects. This prompted them to ensure such education for their sons. Exposure to a liberal education inspired them to think freely and question old ideas and beliefs. Thus gradually a section of the middle class was transformed into the intelligentsia. However, the intelligentsia did not wholly belong to the middle class. A substantial section belonged to the propertied aristocracy—to families that had made their wealth mainly from land and trade. The administrative requirements of the government led to the whole educational machinery being geared to satisfy the needs of public service. This perpetuated the old emphasis on literary education as a virtual monopoly of the upper castes of Hindu society.

J.H. Broomfield while applying the 'elite' concept equates the term with a local term *bhadralok* and known only in Bengal, viz., the 'respectable men'.[81] It was, he says, a socially privileged and consciously superior group; economically dependent upon rental income from land and professional and clerical employment who kept its distance from the masses by its 'acceptance of high caste prescriptions' and its command over education and continuously tried to extend its social powers and political opportunities. It was a status group. Anil Seal shares with Broomfield the non-class conceptual theme.[82] Sabyasachi Bhattacharya, however, argues that though the *bhadralok* did not constitute a class it does not mean that we cannot locate them in a class framework.[83]

Sumit Sarkar believes that an analysis of the ideas and socio-economic roots of the nineteenth-century intelligentsia reveals a significant contrast between 'broadly bourgeois ideals derived from a growing awareness of contemporary developments in the West, and a predominantly non-bourgeois social base'.[84] They 'diligently cultivated the self-image of a "middle class" (*madhyabitta-sreni*), below the zamindars but above the toilers'.[85] It searched for its model in the European 'middle-class' that had ushered in the modern age through movements, viz., Renaissance, Reformation, Enlightenment and the democratic revolutions and reforms. Yet, its own social roots lay not in industry or trade but in government service or the professions of law, education, journalism or medicine with which was very often combined some connection with land.

Considerations for material benefits provided the middle class with the highest incentive for Western education, which provided opportunities for social upgradation. With the middle class intelligentsia gradually becoming significant through jobs and professions, the value of English language education increased. Western education produced both a liberal, progressive and moderate intelligentsia with command over classical learning and keen for reforms as also a more radical pro-reform section that was quick to imbibe the West. If Rammohun Roy was the best representative of the former then Young Bengal was certainly that of the latter. However, the nineteenth-century literati did not only consist of persons brought up in institutions where knowledge was imparted through the medium of English language.[86] Traditional education also produced men who were part of intelligentsia society.[87] These included men who did not deny the importance of Western education and knowledge. They were at the same time moderate and liberal in their outlook and equally keen to reform their society. In fact, there was little to choose between them and the Western educated moderate liberals. Ishwarchandra Vidyasagar is the most outstanding example of a liberal and progressive mind with a classical education.[88] Through his efforts he imbibed the best in the Western system. He shared with the Western Humanists, a tremendous love for humanity and devoted his entire life to alleviate the sufferings of the less fortunate. A like-minded younger contemporary was Harinath Majumdar, popularly called Kangal Harinath, who edited *Grambarta Prakashika*. For a man who spent his entire life in the rural countryside and had no formal training in Western education his understanding of Bengal's land system and the existing agrarian relations was far better than that of many of his Western educated contemporaries and his sympathy for the suffering peasantry more sincere.

The Bengali literati were thus far from a homogeneous group. Its members came from the propertied classes as also from the middle and lower income groups and generally from the upper castes. It consisted of those educated in traditional philosophy and sciences and in the vernacular; the Western educated and; also those who had no formal training in Western education but acquired such knowledge at their own initiative. A significant fact is that Muslims were conspicuous by their absence from the new Bengali literati till the 1860s.[89] In the first half of the nineteenth century, those well-versed in Islamic law and religion constituted the dominant section within the Muslim intelligentsia. Among them were Shariatullah, Titu Mir and Dudu Mian. Between 1818 and 1870 socio-religious movements, viz., Farazi and Wahhabi movements spread over the countryside. However, their influence gradually waned. Gradual acceptance of British rule and Western education saw the emergence of a Western-educated generation.[90] By the end of the century, the Muslim literati comprised a group of traditionally educated men who favoured Western education. They wanted a better deal for the Muslims within the existing politico-administrative framework. There also emerged a group of Western-educated men. Along with these two groups the traditionally vernacular educated section continued to be influential. Some of them were prolific writers who using different literary forms frequently expressed their views on pertinent issues.

MAJOR INFLUENCES ON THE LITERATI

In the nineteenth century various factors influenced the mental make-up of the gradually emerging intelligentsia. It was difficult for them to remain untouched by oriental influence. It was the introduction of English education which marked a turning point in mental emancipation and rationalism, the essential attributes of the educated. It became possible for them to get better acquainted with Western philosophy, thought and science. This enabled them to become interpreters of change in the country.

The traditional milieu, initial training of some of the members of the intelligentsia at the *pathshala* and the steady emergence of an indigenous intellectual ambience prevented a total alienation of the intelligentsia from its traditional cultural heritage. When socio-religious reform was initiated and later when Hindu society and religion was subjected to missionary attack, pride in their traditional identity was awakened. This manifested itself in their study of ancient history and its application to serve definite purpose. Indian cultural and intellectual tradition was sponsored in the 1840s and 1850s more concertedly and actively by the Tattvabodhini Sabha, a forum of both the traditionalists and reformists. Under the aegis of the Sabha, attempts were made to mobilize public opinion in defence of Indian religion and culture against the attacks of the Christian missionaries. The Sabha sponsored the study of Indian history and culture. Rajnarayan Bose, for instance, translated the Upanishads in English, while Akshay Kumar Dutta wrote *Bharatbarshiya Upasak Sampraday.*[91]

Traders, planters, Christian missionaries and others from Europe, called 'interlopers' by the Company's servants, also provided intellectual stimulation. Though representing different shades of opinion, they were unanimous about the degenerate condition of India. In their own way they prescribed that India's emancipation lay in her adoption of Western culture embodied in Christianity, modern education and British institutions.[92] It is however, significant that the attack of the missionaries on Indian religion and culture strengthened the resolve of some of the highly educated men, viz., Rammohun Roy, to free them from the existing social evils. The traders and planters championed unobstructed freedom of trade, the opening up of Indian market to British manufacturers, and the right of the British to settle as well as own property in India. They thus represented a powerful voice of dissent and opposition, attracting conscious Indians, many of whom cherished grievances against the British administration. They were able to portray themselves as philanthropists and friends of Indians. They not only built professional connections with the Indian business community, but also looked to greater cultural and political collaboration by founding Indo-European societies and associating Indians with their movements. Free traders founded the British Indian Society in July 1839. Through trade and service to the Company many Bengalis built-up great fortunes. Free Trade orientation of the Bengali intelligentsia that reached its culmination with the advent of George Thompson in 1843 thus provided intellectual succour and material oppurtunities.[93] Not only did he advise the intelligentsia not to lose faith in England's love for justice and its concern for the cause of the poor but also instructed educated Bengalis to ultimately appeal to the British Parliament.[94] It was English education which greatly moulded the minds of the Bengali intelligentsia.[95]

Among Western schools of thinking, the Utilitarians had a profound impact.[96] The ideas of Bentham and Mill became a dominant influence on Indian administration for quite some time.[97] Utilitarianism with its concept of the greatest good of the greatest number, individualism and good government, evoked favourable responses from the intelligentsia. Bankimchandra Chattopadhyay for instance, was influenced by Bentham and Mill's humanism and Rousseau's concept of equality.[98] He propagated these ideas through the *Bangadarshan* by writing articles on equality and on J.S. Mill.[99] Influenced by the utilitarian philosophy members of the intelligentsia learnt to look to the British rule from the point of view of its usefulness.[100] They appreciated it for its material benefits and deprecated it increasingly for its limitations. They might have imbibed the idea of good government through land, legal and administrative reforms as would be capable of bringing welfare for the greatest number. Similarly pro-peasant stance of some members of the educated middle class, particularly of Bankimchandra possibly exhibits utilitarian impact. Moreover, some of them like Rammohun supported social reforms through legislation as the utilitarians propounded. But the entire philosophy of utilitarian thinkers could not sway the educated section. Moreover, the concept of simple good government without any prospect of self-government and authoritarianism that it came to stand for, naturally caused disillusionment.[101] The utilitarian involvement of the Bengal intelligentsia waned also because the concept became a spent force in England itself after the 1850s and the ruler-ruled dichotomy came increasingly to the fore during the second half of the nineteenth century.

In the 1870s and 1880s positivism replaced utilitarianism as an inspirer of the intelligentsia. Inspired by the ideas of Comte and Congreve, they considered it their duty to realize their idea of humanity in their lives.[102] The anti-colonial and anti-imperialistic content of positivism appealed to them and particularly after anti-Ilbert Bill agitation,[103] they became more alive to its relevance. The educated community was also influenced by Spencer's secularism and writings of Hamilton, Emerson and Theodore Parker.[104] The educated section thus became intellectually dependent on the West. During this period the Liberal Party of England also exerted great influence on the educated middle class. While in England for higher education many were inspired by Fawcett, Gladstone, Burke and John Bright.[105] The extension of franchise, the growth of elementary education, the opening up of civil service and compulsory examination, granting legal status for the Trade Union and the Gladstonian advocacy of Home Rule for Ireland, which the Liberal politicians had been championing were attractive to the intelligentsia.

During the first quarter of the twentieth-century developments within and without affected the thought process of the intelligentsia. The Partition of Bengal, the Morley-Minto Reforms, the First World

War and its aftermath brought in its wake disillusionment and dissatisfaction. Influx of Western thinking gave new dimension to the thought process of the intelligentsia. The repressive measures of the imperial rulers, together with economic degradation alienated a section of the intelligentsia. Their struggle assumed two forms—one was to fight the political and economic stranglehold through the national movement and the other was the expression of the new thinking in literature which would act as a vehicle of progress and constructive work of rebuilding native society and economy. The Russian Revolution and the popularity of Russian literature through translations gave impetus to socialist ideas in Bengal. There was an earnest effort to discover the sorrows and anxieties of the common man. Though no attempt was made to preach a change of the social base, there was a new awareness about some of the social and economic maladies. Social inequalities became the main target.

There was considerable advancement in Western education and those so educated predominated. There was greater preference for liberal professions than for commercial activities. The link between land and intelligentsia weakened considerably. Educated and professional men appeared to be less dependent on land. Government service and the bar continued to attract a considerable section of the educated community. The teaching profession was no less important as also journalism. Some acquired fame as litterateurs. Prominent literati also existed among Muslims. There were eminent literary figures and journalists. Some were actively involved in political activities while some others became members of the legislature.

INTEREST IN AGRARIAN ISSUES

Transformation of Bengal's agrarian economy and the countryside profoundly informed their activities. The intelligentsia reacted to these agrarian issues at two levels. One was at the level of ideas through essays, letters, books and all other forms of literary expressions. The other was at the level of actions, which included among other things, a coherent presentation of their views in the Council, associations, organization of ryot sabhas, etc.

Agrarian thinking of the intelligentsia may be divided into three broad phases—early nineteenth century, late nineteenth century and early twentieth century. The beginning of systematic agrarian thinking can be traced back to Rammohun Roy. A significant section of the early nineteenth-century intelligentsia gave evidence of pro-ryot sympathies. Widespread peasant unrest further attracted intelligentsia's attention to agrarian issues.

The Rent Act of 1859 and the question of peasant right, soon divided the intelligentsia into two factions—pro-landlord and pro-ryot. The reactions of the intelligentsia to agrarian issues were not homogeneous. There was a growing demand for legislative interference to remove all anomalies. Agrarian thinking however, was not restricted to the question of land rights. Many concentrated on problems related specifically to agriculture and its development. Many members of the intelligentsia had first hand knowledge of agriculture. With the passage of time the interest of the intelligentsia in agrarian issues increased further.

In the twentieth century agrarian thinking reached a level of maturity. Literary representation was frequent finding reflections in journals, newspapers and other literary forms. Attempts were made by a section of the erstwhile literati to mobilize peasants on issues relevant to them. Yet there was also a conscious attempt by a section to steer clear of those divisive agrarian issues that could disturb the united fabric of Indian society that they tried to portray before the imperialist forces.

NOTES

1. B.B. Chaudhuri, 'Agrarian Relations, Eastern India', in Dharma Kumar (ed.), *The Cambridge Economic History of India*, vol. 2, *c. 1757-c.1970*, Orient Longman in association with Cambridge University Press, Delhi, 1984, p. 89.
2. *Report on the Land Revenue Commission, Bengal* (1940), vol. I, p. 18.
3. Partha Chatterjee, *Bengal 1920-1947*, vol. I: *The Land Question*, K.P. Bagchi & Co., Calcutta, 1984, p. 14.
4. *Sambad Prabhakar*, 5 Bhadra, 1264 (B.S.).
5. Regulation 7 or Haftam of 1799, also known as the Law of Distraint, permitted landlords to distraint crops for arrears of rent and compel the attendance of tenants at their courts.
   Regulation 5 or Panjam of 1812, also known as the Law of Eviction, further empowered the zamindars to make whatever terms they wished, regardless of custom and, if necessary, by evicting old tenants and settling new ones.
6. B.B. Chaudhuri 'Agrarian Economy and Agrarian Relations in Bengal, 1859-1885', in N.K. Sinha (ed.), *The History of Bengal, 1757-1905*, Calcutta University Press, Calcutta, 1967, p. 305.
7. Radharaman Mukherjee, *Occupancy Rights: Its History and Incidents*, University of Calcutta, Calcutta, 1919, p. 60.

8. Chaudhuri, 'Agrarian Economy and Agrarian Relations in Bengal, 1859-1885', op. cit., p. 276.
9. Mukherjee, op. cit.
10. K.K. Sengupta, *Pabna Disturbances and the Politics of Rent, 1873-1885*, People's Publishing House, New Delhi, 1974, p. 117.
11. Bengal Land Revenue (Misc.) Progs., September 1881, Board of Revenue to the Government of Bengal, Revenue Department, 3 June 1881, para II.
12. B.B. Chaudhuri, 'The Agrarian Question in Bengal and the Government, 1850-1900', in *Calcutta Historical Journal*, vol. I, no. 1, July 1976.
13. *The Gazette of India, Extraordinary*, 3 March 1883, pp. 129-30.
14. Sugato Bose, *The New Cambridge History of India, Peasant Labour and Colonial Capital: Rural Bengal Since 1770*, Cambridge University Press, Cambridge, 1993, Chapter 3.
15. J.C. Jack, *Faridpur Settlement Report* (*1904-14*), p. 29.
16. Bose, op. cit., p. 91.
17. Mukul Mukherjee, 'Impact of Modernization on Women's Occupation: A Case Study of the Rice-Husking Industry of Bengal', *IESHR*, vol. 20, no. 1, 1983, pp. 27-45.
18. Bose, op. cit., p. 8.
19. J.C. Jack, *Bakharganj Settlement Report, 1900-08*, Calcutta, 1915, pp. 10-11.
20. Bose, op. cit.
21. Around 1920 it accounted for 34 per cent of the net cropped area in East Bengal and 18 per cent in West Bengal.
22. Birendranath Ganguli, *Trends of Agriculture and Population in the Ganga Valley: A Study in Agricultural Economics*, Metheun & Co. Ltd., London, 1938, p. 244.
23. George Blyn, *Agricultural Trends in India, 1891-1947: Output, Availability and Productivity*, University of Pennsylvania Press, Pennsylvania, 1966, pp. 102-7.
24. Ira Klein, 'Malaria and Mortality in Bengal, 1840-1921', *IESHR*, vol. 9, no. 1, 1972; and B.B. Chaudhuri, 'Agricultural Production in Bengal: Co-existence of Decline and Growth', *Bengal Past and Present*, vol. 88, no. 2 (166), July-December 1969.
25. Bose, op. cit., pp. 26-7.
26. B.B. Chaudhuri, 'Agricultural Growth in Bengal and Bihar: Growth of Cultivation Since the Famine of 1770', in *Bengal Past and Present*, 95, 1976, pp. 172-3.
27. Bose, op. cit., pp. 28-9.
28. GOI, *Census of India 1921*, vol. V: Bengal, pt. I, p. 389 cited in Bose, ibid.
29. Daniel and Alice Thorner, *Land and Labour in India*, Asia Publishing House, Bombay, 1962.
30. Bose, op. cit., pp. 38-40.

31. Ibid.
32. B.B. Chaudhuri, 'Growth of Commercial Agriculture in Bengal 1659-1885', *IESHR*, vol. 7, no. 1, March 1970, pp. 25-60.
33. Ibid.
34. Ibid.
35. Bose, op. cit., p. 42.
36. Ranajit Guha, 'Neel Darpan: The Image of a Peasant Revolt in a Liberal Mirror', in *Journal of Peasant Studies*, vol. 2, no. 1, 1974, pp. 1-46.
37. Cited in B.B. Chaudhuri, *The Growth of Commercial Agriculture in Bengal*, Indian Studies Past and Present, Calcutta, 1964, p. 80.
38. Raja Rammohun Roy, *English Works*, ed. K. Nag and D. Burman, Sadharan Brahmo Samaj, Calcutta, 1947, vol. IV, p. 83.
39. Chaudhuri, *The Growth of Commercial Agriculture in Bengal*, op. cit., p. 133.
40. Bose, op. cit., p. 51.
41. *Hindoo Patriot*, 25 December 1857.
42. Bose, op. cit.
43. Ibid.
44. Chaudhuri, 'Growth of Commercial Agriculture in Bengal: 1859-1885', op. cit., p. 45.
45. Gopal Chandra Das, *Report on the Agricultural Statistics of Rangpur*, Calcutta, 1874, para 19.
46. Chaudhuri, 'Growth of Commercial Agriculture in Bengal: 1859-1885', op. cit., p. 47.
47. Chaudhuri, 'Growth of Commercial Agriculture', *IESHR*, vol. 7, no. 2, 1970, pp. 240-1.
48. Rajat K. Ray, 'The Crisis of Bengal Agriculture-Dynamics of Immobility', *IESHR*, vol. 10, no. 3 (1973), pp. 261-2.
49. Cited in B.B. Chaudhuri, 'Agrarian Relations: Eastern India', in Dharma Kumar (ed.), *Cambridge Economic History*, vol. II, *c. 1757-1970*, op. cit., p. 147.
50. Bose, op. cit., pp. 64-5.
51. Blyn, op. cit., p. 96.
52. Bhaduri, op. cit.
53. Bose, op. cit., pp. 123-4.
54. Ibid.
55. J.C. Jack, *Faridpur Settlement Report (1904-1914)*, pp. 30-1.
56. Amit Bhaduri, 'A Study of Agricultural Backwardness under Semi-feudalism', *The Economic Journal*, 83, March 1973.
57. Chaudhuri, 'Agrarian Economy and Agrarian Relations in Bengal, 1859-1885', in N.K. Sinha (ed.), op. cit., pp. 317-18.
58. *Hindu Ranjika*, 9 February 1875 (RNP).
59. Chaudhuri, *The Growth of Commercial Agriculture in Bengal*, op. cit., p. 159.

60. Cited in ibid., p. 202.
61. Sengupta, *Pabna Disturbances and the Politics of Rent, 1873-1885*, op. cit.; 'Agrarian Disturbances in the Nineteenth Century Bengal', in *IESHR*, 8, 1972, pp. 192-212.
62. Bose, op. cit., p. 157.
63. Sengupta, *Pabna Disturbances*, op. cit., p. vii.
64. B.B. Chaudhuri, 'The Transformation of Rural Project in Eastern India, 1757-1930' (Presidential Address, Modern Indian History Section, 40th Session, Indian History Congress, Waltair, December 1979), p. 37, cited in Bose, op. cit., p. 159.
65. 'Report on the Origin, Nature, and Probable Consequences of Agrarian Combinations in East Bengal' from J.G. Charles, dated 19 September 1873, Misc. Colln. 14, nos. 26-7, Bengal Land Revenue Progs, January 1874, cited in Sengupta, *Pabna Disturbances*, op. cit.
66. GOI, *Census of India, 1921*, vol. V. Bengal, pt. I.
67. Bose, op. cit., p. 32.
68. Rajat Ray, 'The Crisis of Bengal Agriculture: Dynamics of Immobility', in *IESHR*, vol. 10, no. 3, 1973, pp. 261-2.
69. Bose, op. cit., p. 57.
70. Ray, 'The Crisis of Bengal Agriculture', op. cit., pp. 261-2.
71. Amiya Bagchi, *Private Investment in India, 1900-1939*, Cambridge University Press, Cambridge, 1972, p. 287.
72. Bose, op. cit., p. 59.
73. Cited in Chaudhuri, 'Agrarian Relations: Eastern India', in Kumar (ed.), *Cambridge Economic History*, vol. II, *c. 1757-1970*, op. cit., p. 147.
74. J.C. Jack, *Economic Life of a Bengal District*, Oxford, 1916, pp. 81-2.
75. Bose, op. cit., p. 86.
76. Ibid., p. 90.
77. *Report of the Land Revenue Commission, 1940*, vol. I, p. 20.
78. Sachin Sen, *Studies in the Land Economics of Bengal*, 1935, p. 281.
79. B.B. Misra, *The Indian Middle Classes: Their Growth in Modern Times*, Oxford University Press, Delhi, 1978.
80. *Grantidar* was a form of intermediate tenure in Jessore-Nadia region.
81. J.H. Broomfield, *Elite Conflict in a Plural Society: Twentieth Century Bengel*, University of California Press, Berkeley, 1968, p. 5.
82. Anil Seal, *The Emergence of Indian Nationalism: Competition and Collaboration in the Late Nineteenth Century*, Cambridge University Press, Cambridge, 1968, p. 341.
83. Sabyasachi Bhattacharya, 'Notes on the Role of the Intelligentsia in Colonial Society: India from Mid-Nineteenth Century', *Studies in History*, vol. 1, no. 1, January-June 1979, pp. 93-8.
84. Sumit Sarkar, *Modern India, 1855-1947*, Macmillan, New Delhi, 1983, p. 67.
85. Ibid.

86. The literacy figures even in 1911 were only 1 per cent for English and 6 per cent for the vernaculars.
87. Benoy Ghosh, *Banglar Samajik Itihaser Dhara, 1800-1900* (Bengali), Ayan Publishing, Calcutta, 1968, p. 138.
88. Asok Sen, *Ishwarchandra Vidyasagar and his Elusive Milestones*, Riddhi, Calcutta, 1977.
89. Benoy Ghosh, *Banglar Biddyotsamaj* (Bengali), Ayan Publishing, Calcutta, 1973, p. 22.
90. Rafiuddin Ahmed, *The Bengal Muslims 1841-1906: A Quest for Identity*, Oxford University Press, Delhi, 1981, pp. 136-7.
91. D.K. Biswas, 'Tattvabodhini Sabha O Debendranath Thakur', *Itihas*, vol. V, no. 3, pp. 164-9.
92. Eric Stokes, *The English Utilitarians and India*, Clarendon Press, Oxford, 1959, pp. 34, 54-5.
93. Sibnath Sastri, *Ramtanu Lahiri O Tatkalin Bangasamaj* (Bengali), New Age Publishers, Calcutta, 1955, p. 153.
94. S.R. Mehrotra, *The Emergence of the Indian National Congress*, Vikas, Delhi, 1971, pp. 23-5.
95. *Calcutta Monthly Journal*, March 1837, pp. 82-3.
96. Stokes, op. cit., p. XII.
97. Ibid., p. 161.
98. Sastri, op. cit., p. 253.
99. *Bangadarshan*, Kartik B.S. 1282 (1875) and Poush B.S. 1284 (1877).
100. B.B. Majumdar, *History of Indian Social and Political Ideas: From Rammohun to Dayananda*, Bookland Pvt. Ltd., Calcutta, 1967, p. 85.
101. Stokes, op. cit., pp. 57-79.
102. B.B. Gupta, *Puratan Prasanga* (Bengali), Vidyabharati Sanskaran, Calcutta, B.S. 1966, pp. 11, 31, 41.
103. In the 1880s there was a vehement protest by the British against Viceroy Ilbert's attempt to introduce a bill.
104. S. Sastri, *Atmacharit* (Bengali), Sadharan Brahmo Samaj, Calcutta, rpt., 1982, p. 92.
105. Jogesh Chandra Bagal, *Muktir Sandhane Bharat: Congress Purva Yuga* (Bengali), Bharati Library, Calcutta, 1972, pp. 129-30; S.N. Banerjee, *A Nation in Making: Being My Reminiscences of Fifty Years of Public Life*, 1925, rpt.; Oxford University Press, Bombay, 1966, p. 36.

CHAPTER 2

Evolution of Agrarian Thinking

Giyachhe punjipata, bhitete shakul kanta
Amar dhan giyachhe maan giyachhe
Ekhon ma pran niye sangshay
Gelo goru, trina toru, kichu nai aar[1]

ISHWARCHANDRA GUPTA

The beginning of systematic agrarian thinking in Bengal can be traced back to Raja Rammohun Roy. Long before the beginning of 'economic nationalism' best represented in the writings of Dadabhai Naoroji, M.G. Ranade and Romesh Chandra Dutt, Rammohun pioneered a trend of thinking that attempted to analyse the structural changes that were introduced in the economy in the wake of colonization. He was perhaps the best representative of the early generation of the modern Bengali intelligentsia that stood at the crossroads of a process of economic change which, had been initiated half a century ago.

THE DAWN OF A NEW AGE

Even while representing a generation that owed its economic well-being and sense of identity to the advantages offered by the colonial economy he attempted a critique of the impact of land legislations introduced by the British on the rural society of Bengal. Rammohun's father in the early years of the nineteenth century had acquired a zamindari when many zamindari estates were being sold at public auctions, after the introduction of Permanent Settlement in 1793. He himself added to his patrimony by service under the Company. His direct experience of estate management and his knowledge of classical and Western economic and philosophical thought gave him a rare insight into the workings of the colonial land system, problems of European Settlements and the monopoly business of the East India Company. Along with his interest in the *sastras*, contemporary problems of politics and economics informed his thinking.[2] In analysing the framework of his thought it must be noted that by birth he belonged to the landed gentry and that he himself made money out of usury.

Rammohun's perception of the colonial impact on the rural society and economy of Bengal is documented in his reply to the questionnaire sent to him among others, by the Select Committee appointed by the British House of Commons in 1831 to examine the question of the renewal of the Company's Charter due in 1833. His perception of the evolution and consequences of the land revenue system is explained in his answers to the fifty-four questions and in the long article appended to this evidence to clarify some of his points furnished in the text of his evidence. He also answered the thirteen cognate questions framed by the Committee of which six were directly related to the Indian peasantry and their livelihood.

Rammohun attempted the first appraisal of the Permanent Settlement and its consequent effects on the agrarian economy of Bengal.[3] He examined at length the impact of it on the condition of the two dominant sections of agrarian society—landlords and ryots. The former he concluded had benefited immensely from the Settlement, having been recognized as the sole class with an unqualified proprietary right in the soil. The ryots were denied any such rights. Ignoring all attempts to restrict rent beyond that estimated half the gross produce of the land fixed in theory, zamindars were able to enhance rents much beyond that amount by various means. In his understanding, the government had declared in Regulation VII of 1793 that the *pattahs* fixing the rates of payment for the lands of resident or *khud-kasht* ryots could be generally cancelled only on four grounds—if they had been obtained by collusion; if the rents paid by the *khud-kasht* ryots within the last three years had been below the *nirkhbundee* (general rate) of the *purgunnah*, i.e. particular part of the district where the land was situated; if the *khud-kasht* ryots had obtained collusive deductions or; upon a general measurement of the *purgunnah* for the purpose of equalizing and correcting the assessment. But in actuality Rammohun observed, the rents of all types of ryots not excluding the *khud-kasht* peasants could be enhanced arbitrarily by zamindars in connivance with the local revenue officials. In Bengal, he remarked, 'the landlords have met with indulgence from government in the assessment of their revenue while no part of this indulgence is extended towards the poor cultivators' who were left totally at the mercy of the zamindar's avarice and ambition.[4] The latter effectively exacted in addition to the revenue an even higher rate through all kinds of extra-legal means at their disposal. It was left to the struggling peasant to bear the cost of the cultivation including that of livestock, cattle and seeds.

Rammohun questioned the principles that underlay the government's attempts to fix the revenue. In Bengal at the time of the Permanent Settlement, the amount of the revenue, which had been paid by each zamindari in the preceding year, was taken as a standard of assessment, subject to certain modifications. *Taluks* which had paid a revenue directly to the government for the previous twelve years without any fluctuation, were to be assessed at that rate, and the principle of that assessment was considered to be nearly one half of the gross produce. The different fields or plots of land on an estate were classed into four categories depending on their quality and certain rates per *bigha* were affixed to them respectively, agreeable to the established rates in the district. These rates were considered as a standard in settling the rent to be paid by the cultivators. But in spite of all these provisions made in the settlement Rammohun observed the precise quality of land was always liable to dispute. Land might be classed in the first, second, third or fourth quality according to the discretion of the zamindari or government surveyors. Measurement was always liable to variation through the 'ignorance, ill-will, or intentional errors of the measurers—there is "in practice" no fixed rate or amount of rent demandable from them, although such a standard is laid down in theory'.[5] Rent was generally paid in money. But modernization in the sense of the introduction of the system of cash rents, according to him, had not made much difference to the chronic poverty of the agricultural tenant. Money rent was usually paid in monthly installments, the heaviest payments being made when the harvest was realized.

Rammohun admitted that the landlord gained as a class. In an abundant season, when the price of corn was low, the peasant was required to sell the major portion of his crop to meet the demands of the landholder leaving little or nothing for seed or subsistence to the labourer or his family. Rammohun correctly observed that landlords could improve their lot further, because of the fact that under the terms of the Settlement the state was to collect from them a fixed total amount and not a fixed proportion of rent. This total was not linked to the size of the landlords' holding. Nor was it dependent on the total production. Hence if the landlord could bring fallows under cultivation then it would be practically free of any rent.

The benefits which the landlords enjoyed was thus due to the extended cultivation of waste lands which formerly yielded no rent and the subsequent increase of rents much beyond those rates paid by cultivators at the time of the permanent settlement. While Rammohun did not disapprove of the augmentation of the income of zamindars as a result of extension of cultivation he was rather critical of the latter.

He was convinced of the fact that landlords had failed to live up to the expectations of the framers of the Permanent Settlement. They had failed to emerge as a class of entrepreneurs or bring about much improvement in their lands and cultivation. No agricultural revolution of the English type happened in Bengal. On the contrary, change in the composition of the landed class with investment of trading capital in land facilitated by sale of estates at least in the initial years, resulted in zamindari management being entrusted to a growing class of intermediaries who were just parasites living off the fruits of toil of cultivators.

As to the benefits that accrued to the state from the settlement, Rammohun observed that the state did not lose financially. His argument was that the rate fixed was higher than any hitherto fixed rates and the government sacrificed nothing in concluding the Settlement. He did not realize that the state gave up for all times any opportunity of having the rates enhanced. Since land revenue was the principal source of government revenue and since its total amount became permanently fixed, it necessarily followed that to meet additional expenses and other contingencies the state would have to levy new taxes. This meant that while the landlords as a class benefited the other sections of the society had to bear the burden of additional taxes.

Rammohun penned a dispassionate account of the fate of ryots under the changed circumstances. He recognized their deplorable condition, which was accentuated by the high rental demand. Though he referred to the absolute right enjoyed by *khud-kasht* ryots to continue possession of their lands in perpetuity under earlier regimes, he did not raise the question of occupancy right. The regulations of 1793 deprived the ryots, including the *khud-kasht* ryots, of their occupancy rights for the first time. It was prescribed that he would cultivate the land; expenses on account of livestock, cattle and seeds would be borne by him and; half of the produce he would pay to the landlord, the owner of the land, as rent and not revenue. The rest he would keep for himself.

He, however, admitted the failure of ryots to accumulate capital, which could have enriched agriculture in the end if ploughed into land. He explained the failure of the peasants to accumulate capital thus:

> Very often when grain is abundant, and therefore cheap, they are obliged ... to sell their whole produce to satisfy the demands of their landlords, and to subsist themselves by their own labour. In scarce and dear years they may be able to retain some portion of the crop to form a part of their subsistence, but by no means enough for the whole. In short, such is the melancholy condition of the agricultural labourers that it always gives me the greatest pain to allude to it.[6]

Rammohun drew attention to the fact that the proprietors distrained the moveable property of defaulting tenants with the assistance of the local police officers. All these precluded accumulation of capital by the ryots. Consequently, no true economic development was possible. Besides, the zamindars were in the habit of farming out their estates to intermediaries. These intermediaries were much less merciful in their collection of rents. Besides, the courts were often situated far away and the cultivators too poor to undertake the hazardous and expensive enterprise of seeking legal redress. Rammohun stated categorically:

> The power of imposing new leases and rents, given to the proprietors by Reg. I and VIII of 1793, and subsequent Regulations, has considerably enriched, comparatively, a few individuals, the proprietors of land, to the extreme disadvantage or rather ruin of millions of their tenants, and it is productive of no advantage to government.[7]

In his understanding, the government's measurement operations, lands rights and obligations were all instrumental in oppression of peasants. Commenting upon the chronic poverty of this vital section of agrarian society, he observed:

> It soon became evident to anyone who had toured the provinces that within a circle of a hundred miles in any part of the country, there was very few, if any besides the proprietors of land who had the least pretension to wealth or independence, or even the common comforts of life.[8]

He was, however, silent on the fact that the Permanent Settlement had left the landlord and not the actual cultivator as the owners of the land. He accepted the cancellation of the ancient rights of the ryots. Right of ownership of land meant the right to inherit and transfer their holdings. This right was effectively shifted from the cultivator to the landlord by the Permanent Settlement. The far-reaching consequences of this escaped him.

Rammohun was critical of some of the provisions of the Settlement. He understood that the rate of rent was rather exorbitant. This rate was one-half of the total produce, the cost of cultivation and of seeds being borne by the cultivator himself. In addition, the landlords effectively exacted an even higher rate from the cultivators through all sorts of means available at their disposal. The cultivator was forced by influential landlords to cultivate inferior lands at higher rates. Besides, he admitted, the zamindars themselves suffered when their estates were sold for failure to pay the revenue. Often the notices of sales were not publicized. Native revenue officers often had the opportunity of effecting purchases of the land at a reduced price. Since the notices of sale were published in the government gazette, the proprietor often was unaware of it. In cases where the agents of the zamindar conducted the business, the former often with the help of the revenue officers sold the land at a very low price. Rammohun suggested that the advertisements or notices of sale should first be regularly sent to the parties interested, at their own residences, not merely delivered to their agents. They should be fixed up not only in the government offices, but at the chief marketplaces and ferry *ghats* of the district and in those of the principal towns. The police officers should be required to take care that the notices remained fixed up in all these places from the first announcement until the period of sale. The date and time of sale being precisely fixed, the bidding for an estate should be allowed to go on for a specific period to enable all intending purchasers to make an offer. This indicates that while he objected to the unlimited power given to the zamindars to increase the rents of the *khud-kasht* ryots, Rammohun had no desire to see the zamindars exploited.

Rammohun admitted the need for improvement in agrarian relations to relieve the chronic poverty of the common peasant. It was imperative for the latter to be assured that he had a stake in the land as an owner cultivator, or at least, as an occupancy tenant with full security of tenure, who paid a reasonable rent. Once this happened, there was a chance for him to accumulate capital. If, at the same time, he had access to improved technique and other facilities for increased production, which economic modernization could ensure, there was a clear possibility of a massive economic transformation.[9]

Rammohun outlined a comprehensive plan of land reforms. In

fact, his plan was to set the mould of progressive thought on land reforms for the rest of the nineteenth century. Rammohun believed that if the government intended to improve the condition of the peasantry it should absolutely forbid any further increase of rent on any pretext whatsoever. The condition of the Indian peasants was that of 'agricultural labourers' rather than that of 'peasant proprietor'.[10] There should be no further scope for fresh survey and measurement of lands. The adoption of such a measure would be justified in view of the fact that while introducing the permanent settlement[11] the government declared it to be its right and its duty to protect the cultivators 'as being from their situation most helpless', and 'that the landlord should not be entitled to make any objection on this account'. To explain the logic of freezing of rents[12] he argued that during Muslim rule landed proprietors had to pay to the ruler a 'considerable proportion' of the rent collected. The ruler could also increase the revenue rates and even take away the proprietary rights of the landlords when they failed to pay the revenue unjustly alleged to be due from them. Indian landlords had thus been quite unlike English landlords: 'Under these circumstances, the situation of the proprietors was not in any respect on a more favourable footing than that of the khud-kasht tenant, and consequently their right was not in any way analogous to that of a landlord in England.'[13]

Yet the British treated them as English landlords. Rammohun was not averse to the government sparing them any distress and difficulties originating from the uncertainty of assessment. What he found unsatisfactory was the government's failure to protect all types of tenants against the uncertainty of assessment. He said:

> I am at a loss to conceive why this indulgence (shown to the landlords) was not extended to their tenants, by requiring proprietors to follow the example of government in fixing a definite rent to be received from each cultivator, according to the average sum actually collected from him during a given term of years.[14]

He failed to comprehend why the government, out of compassion for the miserable condition of the cultivators, did not fix a 'maximum' standard, corresponding with the sum of rent now paid by each cultivator in the year, and positively forbid any further increase in the rental.

This would not involve any real violation of the regulations of the Permanent Settlement. Regulation I had declared that[15] the government would whenever it deemed proper, enact such regulations necessary, for the protection and welfare of the dependent ryots and other cultivators of the soil. Similarly, Regulation VIII of 1793 provided for the permanent right of a *khud-kasht* tenant to retain possession of land at a 'fixed rent' and protected him against cancellation of his title deeds, except under certain conditions. What Rammohun was proposing was in essence a Permanent Settlement for the ryots. It would be a logical corollary to the Permanent Settlement of the landlord's revenue. Nevertheless, freezing of rents was not enough, because the rent was already raised very high. The high rent rate also hindered provident husbandry and accumulation of capital for farm improvements. Therefore, the ryots' rents should be reduced and stabilized and along with this the revenues payable by the landlords should be lowered. The loss of income that the government would suffer because of the lowering of the revenues payable by the zamindars should be made good by 'taxes on luxuries and such articles of use and consumption, as are not necessaries of life'. This in itself was a remarkably advanced position for Rammohun Roy to have taken in the early nineteenth century. He felt that it was feasible to tap new sources of indirect taxation of luxury goods. Increased demand for agricultural commodities since 1814 had enriched the landlords and the dealers in agricultural commodities. Their luxury consumption as well as that of Europeans could be easily taxed.[16] He also suggested that the large government expenditures of those days were capable of being sharply reduced.[17] An effective method for securing a reduction would be to appoint Indians in place of Europeans in most of the administrative and judicial posts. It is interesting to note that even as early as 1831 he was already speaking of a drain of wealth.

For a general improvement in the condition of the country, Rammohun pleaded for economic modernization. This meant free enterprise. He was well aware that the comparative freedom of trade after 1813 had created an illusion of prosperity. Nevertheless, he maintained that there had been no 'common increase of wealth'. He said that there was an extreme difference which existed between the rate of value at which estates sold prior to the year 1793, or even several years subsequent to that period, and the present.[18] This enormous augmentation of the price of land was partly due to the extensive cultivation of waste lands and to a larger extent due to the rise of rents payable by the cultivators.[19] But the overwhelming poverty throughout the country with the exception of the towns and their vicinity bore testimony to the fact that the increase of wealth in general had not been the cause of the actual rise in the value of the landed estates, in spite of the extensive cultivation of waste lands under the stimulus of the permanent settlement and of freedom of trade. The greatly increased demand for the produce of lands had been attributed to the increased wealth of Bengal, which had been due to the opening of trade in 1814. However, Rammohun observed that the increase of wealth was confined to landlords and dealers in commodities. The government also appropriated an enormous duty on the transit and exportation of the produce of the soil.[20] However, he concluded erroneously that there was considerable increase in the produce occasioned by the extension and the improvement in agriculture by the proprietors, assured of the fact that no demand of an increase of revenue would be made upon them on account of the progressive productiveness of their estates.

What was imperative was agricultural and industrial development with the aid of Western techniques, Western capital and Western enterprise. Economic modernization should involve not just the new Indian middle class but also the impoverished peasants, the masses of the population. The chronic poverty of the agricultural tenant had to be alleviated.

He was firm in his conviction in the need for selective colonization of Europeans in the country. Probably he was assuming that such settlement would mean import of capital and skill and would thus make the economy prosperous. Rammohun reasoned that, since Europeans retired with the fortunes realized in India, a system which would encourage Europeans of capital to become permanent settlers with their families, would necessarily greatly improve the resources of the country.[21] In spite of great opposition[22] he advocated the ownership of real property in India by Europeans.[23] Selective colonization, he was convinced, could initiate a process of economic modernization in India.[24] He said that European settlers in India would introduce the knowledge they possessed of superior methods of cultivating the soil and improving its products. The presence of Europeans would hasten the improvement in the laws and the judicial system. By their very presence in the countryside, they would afford protection to the poor man against 'the impositions and oppression of the landlords and other superiors' and against the official abuse of power.

He was, however, not in favour of unrestricted free trade in respect of the export of food grains from India.[25] He once argued that the export of rice had a tendency to aggravate the effects of the failure of crops and, therefore, on one occasion appealed to the government to suspend free exports of rice from the Bengal ports. Generally, he argued that an export duty on rice could be justified, if the proceeds of the duty could reduce the burden of land revenue, particularly on the tenant farmers.

Rammohun Roy was thus the first thinker to provide a comprehensive critique of the agrarian situation in Bengal. The Permanent Settlement of 1793 was justifiably his starting point as it had initiated a process of change. He said that if the Permanent Settlement had not been introduced, the zamindars would always have taken care to prevent the revenue from increasing by not bringing the waste lands under cultivation and by 'collusive arrangements' to elude further demands; while the state of the cultivators would not have been any better. Therefore, the abolition of the Permanent Settlement was not the answer. The improvement of the lot of the ryots would be possible only if the rents were reduced and the Permanent Settlement was extended to them. He explained that, the introduction of cash rents did not encourage cultivation in a period of rising prices after 1814, because the normal burden of rent on the ryot was excessively heavy. Moreover, since the British had no interest in maintaining or raising the productivity of agriculture, by means of investment in economic overheads like roads, canals or extension of markets, land tax became an instrument of exploitation.

Rammohun, however, failed to realize the true impact of the system of indigo cultivation on the rural countryside. The other eminent Bengali Dwarkanath Tagore shared this shortcoming. This shortcoming becomes even more glaring in view of the fact that the indigo peasants had been agitating against the system for long. Both Rammohun and Dwarkanath were themselves planters in their own right.[26] The latter even held that the poor classes were better off through the diffusion of purchasing power, which was the result of the introduction of a remunerative cash crop.[27] Forced labour exacted by indigenous zamindars was being replaced to some extent by free labour. He observed that the wages of Rs.4 per month paid by the planters was not a miserable income and also that the middle classes employed as *sarkar* and in other capacities in the plantations were getting higher salaries and were no longer at the mercy of the zamindars and the great *baniya*s. The agents of the absentee landlords were far more oppressive. Rammohun agreed that the natives residing in the neighbourhood of indigo plantations were evidently better clothed and better conditioned.[28] Failure to comprehend the inherent exploitative character of the working of the indigo plantation system explained their support to the system in spite of the glaring anomalies.[29]

Rammohun's conscious presentation of the agrarian question brings to light certain major issues, which were to become the focus of a prolonged discussion by the later Bengali intelligentsia. His was the first critique of the Permanent Settlement. The settlement that fixed the revenue at a very high rate was responsible for a corresponding hike in the rent rate. It destroyed gradually the bond of amity between the landlords and the tenants and was detrimental to the cause of economic modernization. The higher rent, the proprietary right granted to the zamindars, their unlimited power to increase the rent due from even the *khud-kasht* peasants, the increasing hostility between the zamindar and the ryots, all contributed towards growing oppression of the latter and their failure to accumulate capital. Rammohun was the first to lay down the principles for amelioration of the condition of the peasants.

THE INHERITORS

The Young Bengal subsequently picked up the threads. The young students of Hindu College in the 1820s and 1830s, because of the unique influence of Derozio came to be clubbed together as a distinct group.[30] Although their paths diverged later,[31] in their early years the Derozians were concerned with religious and social issues.[32] A review of the activities of the Society for the Acquisition of General knowledge bear testimony to this.[33] However, they soon retreated from their early radicalism.[34] The Derozians left no permanent effects in the sphere of religion.[35] In the domain of social reforms too they themselves, failed to organize any real campaign on any issue.

It was only later that some members of this group became conscious about agrarian issues. However, by then almost a decade had elapsed since Rammohun expressed his views before the Select Committee. It is not difficult to explain this early apathy of the Young Bengal towards the problems of the greater multitude of Bengal's population. It could be because they came from the urban classes and even those who had rural roots belonged mainly to the landed classes. It was difficult for them in the 1840s to transcend their class interests and develop awareness for issues not affecting them directly. Age and maturity helped to develop their understanding of these complex issues.

The need for the amelioration of the condition of the ryots was expressed in the columns of the *Bengal Spectator*. In its first issue,[36] it stated that its aim was to educate the people and strive for their happiness. It stressed on the fact that the Bengal ryot had to undergo twofold oppression—oppression by the zamindar and by the government, which legalized the system of zamindari oppression. Speaking about the Permanent Settlement, the journal stated: 'A system such as this country never beheld, was established, by which in the same ratio that government secured the realization of its own revenue, it enabled its farmers and revenue payers to squeeze the last pice from their undertenants.'[37]

The liberal attitude of the journal was revealed in a series of articles titled 'Ryot', which were published by it over a period. By narrating the story of the poor Muslim peasant Miajan[38] of Hooghly, it illustrated how the zamindars and *talukdars* took advantage of the legal loopholes in the land settlement to exploit the ryots. As Miajan soon found out, not only the *talukdar* but also the local police and the collector were involved in oppressing the poor ryots.

At a weekly meeting of the Native Community at Balkhana, Dakshinaranjan Mukherjee, spoke at length on the condition of the ryots under the Hindu, the Muhammedan and the British administrators. He explained that under the native government the right in the soil was not vested in private individuals. It was a great mistake to have converted zamindars, who were the 'collectors of revenue' into proprietors of landed estates, by which act the rights of a vast number had been sacrificed.[39] Dakshinaranjan (1814-78) was related to the branch of the Tagores at Pathurighata. He was born in an orthodox Brahmin family of Bhatpara. The contradiction in the Young Bengal's handling of the peasant problem is revealed in the fact that in later life he settled down in Lucknow where he acquired considerable property. He became a member of the Oudh Talukdar Samiti.

In the columns of the *Bengal Spectator* were narrated at length the evils of the zamindari system. Among these evils were rack-renting, illegal imposts and the oppression of the moneylenders. At a meeting addressed by George Thompson, Ganendranath Tagore exposed the abuses of the zamindari system.[40] At another meeting Dakshinaranjan Mukherjee depreciated the system of middlemen as being highly detrimental to the interest of the ryots.[41] Peary Chand Mitra depicted the zamindars as cruel oppressors of their ryots. He asserted that they imposed illegal taxes of all kinds and resorted to physical torture to gain their nefarious ends. Moreover, there were the moneylenders, who through various indigenous forms of extortion left the cultivators with nothing.[42]

The *Bengal Spectator* vehemently criticized the laws of distraint.[43] The journal held that the distraint laws legalized all kinds of oppression and tyranny by zamindars. Government allowed them to squeeze their ryots, in order that the government could squeeze the zamindars in turn. The summary trials were nothing but occasions for exactions. They were the bane of the Bengal peasants. The journal lamented the fact that, the makers of the Permanent Settlement had paid no thought to the plight of the peasantry. Consequently, there was no end to the misery of the ryots, so much so that the word 'ryot' was synonymous to 'poor'.[44]

With regard to the relation between the landlord and the tenant, the stand of the *Bengal Spectator* was definitely pro-ryot. A correspondent of the journal wrote that the ryots constituted not only the majority of the population but were also the most oppressed.[45] There were occasions when the oppressed peasantry, as a last resort, took to *dharmaghat*.[46] The *Bengal Spectator* advocated self-sufficiency in the sphere of agriculture, trade and commerce.[47] It listed as reasons for the miserable plight of the people of the country—immorality, administrative inefficiency and poverty. The Young Bengal proposed to set up some sort of an organization to carry on constitutional agitation in defence of the peasantry. A committee was set up to find a solution to the major agrarian problems.[48] The committee circulated through the columns of the *Bengal Spectator* a questionnaire, to be filled in by the readers.[49] The questionnaire not only throws light on the oppression of the peasantry in all its aspects, viz., enhancement of rent, illegal imports, the oppression of the moneylenders but also highlights the real condition of the ryots, particularly the loss of their rights. It also sought information on the growing differentiation in the ranks of the peasantry.[50]

Young Bengal leaders highlighted the oppression of the zamindars and the misery of the ryots. More radical than the moderate Rammohun, they were unequivocal in their denunciation of the evils of the zamindari settlement. Another crucial fact, which they emphasized, was the increasing animosity between the landlords and the tenants. They held the Permanent Settlement responsible for the destruction of the 'paternal' relationship between these two classes. In the Young Bengal's attitude towards the major agrarian issues, there was justification of the defensive reactions by the exploited peasantry. Like Rammohun, they had a good idea about the changes taking place in the rural countryside of Bengal over the preceding half century. However, unlike the former, they failed to suggest any comprehensive plan of reform. They failed to suggest a viable alternative. They failed to raise the agrarian question from the conceptual plane to the level of practical politics. Their activities in this sphere were limited to the publication of a few articles in their journal and that too over a very limited period. Besides, only a handful of their leaders were associated with this. In fact, the leaders had no connection with the countryside. They had no real conception about the problems facing the multitude of Bengal's peasantry. Therefore, they could only skirt the surface and offer a half-hearted response. They probably found it fashionable to denounce a class of people who represented elite society, authority and wealth.

In 1833, Rasik Krishna Mullick asserted 'that the only way now to improve the condition of ryots is to effect a reformation in the organization of Mofussil Courts'.[51] Peary Chand Mitra suggested a permanent settlement in rent-rates.[52] He had great faith in the zamindars made benevolent by English education.[53] He appealed to the zamindars: 'When the ryots are well-protected, they find it easier to pay your claims . . . your happiness and the happiness of your ryots, are identified with each other.'[54]

For all their early apparent radicalism at least in their economic thinking, the Derozians gave little proof of anything unique or novel. A basic loyalty towards the British underlies all their ideas.[55] Derozian journals, viz., *Jnananesvana* repeatedly urged their readers to take up independent trade. But they themselves were far from successful in this.[56] Their understanding of the basic economic relationship between Britain and India was handicapped by their own dependence for economic advancement on the British. Thus, even the decline of Indian handicrafts passed unnoticed by them.[57]

SEDATE CONTEMPORARIES

While the Derozians raged their more solid contemporaries', viz., Akshoy Kumar Dutt (1820-86) wielded a more powerful pen. He was a man of wonderful and versatile intellectual interests. Born in a humble family, it was through sheer effort that he educated himself. He soon encountered the poet Ishwar Chandra Gupta, at whose request he began to write regularly for *Sambad Prabhakar*, became a member of the Tattvabodhini Sabha founded by Debendranath Tagore and later adopted Brahmoism.

The members of the Tattvabodhini Sabha founded in 1839 continued the trend of agrarian thinking initiated by Rammohun. In August 1843, they started the publication of the *Tattvabodhini Patrika*, which became their mouthpiece. Until the time when it went out of circulation in 1859, the *Patrika* regularly expressed the views of the Brahmo leaders on major agrarian issues. It sought to adopt a modern economic outlook.[58] While it favoured economic development, it was also quick to condemn all forms of socio-economic exploitation.

The *Tattvabodhini Patrika* in the 1840s published a series of articles, which give an indication of the perception of the agrarian question of the members of the Tattvabodhini Sabha.[59] These articles clearly depict the miserable condition of the Bengal peasantry. The colonial government as well as the local lords oppressed them. The *Patrika* categorically stated that land was their capital and the ryots, their protector. However, it was lamentable that their misery knew no bounds. The reason was the unrestrained oppression of the zamindars. The latter fleeced the ryots, subjecting them to extra-legal and illegal imposts. In some cases, the zamindars increased the dues of the ryots by one fourth of the total unpaid dues of the ryots. The *Patrika* gave an instance of how the zamindars increased their dues. A Calcutta-based Brahmin zamindar once asked his peasants for a handful of grains as alms. Then he calculated that this collection amounted to Rs.1,500. He then declared that henceforth the ryots would have to pay in addition to their other dues a yearly sum of Rs.1,500. This was just one mode of oppression. Each zamindar devised different methods for oppressing the ryots.

Added to the oppression of the zamindars was the exploitation by their agents and servants. The latter collected their dues from the ryots on their own. The misery of the ryots became complete if the zamindar leased out portions of his land to the *ijaradars* or revenue farmers. Unlike the zamindars, the revenue farmers had no stake in the land. While the former did not desire the absolute ruin of the peasants for fear of ruining agriculture, which would prove detrimental to his interests in the end, the latter had no such compunction. His dues to the zamindar or the superior leaseholder were fixed. Anything he collected over and above this due went into his own pockets. Since the term of his lease was fixed, he had no interest in the improvement of the land. His sole interest was to get rich as quickly as possible at the expense of the ryots. The ryots thus had to serve four masters: the zamindar, the *pattanidar*, the *ijaradar*, and the *dar-ijaradar*.[60]

The *Patrika* traced back the real source of the ruin of the ryots to the Permanent Settlement. Cornwallis's dream that the Permanent Settlement would improve the condition of the rural economy had not been realized. The revenue had been fixed at a very high rate and was rigorously collected. The zamindars had to pay the revenue by the sunset of a specified date, irrespective of the fact whether the ryots could pay or not. Therefore, the oppression of the ryot, left totally at the mercy of the zamindar, was the logical corollary. The *Tattvabodhini Patrika*[61] described how the zamindars exercised their rights over not only the moveable and immoveable property of the ryots but also over the latter's person and his physical labour. His plight was worse than that of a slave. The slave had the right to be looked after by his master, but the ryot had no such right. The *Patrika* listed eighteen different types of physical torture practised upon the hapless ryot by their zamindars.[62]

The ryots had no means of redressal. There were law courts but they were often situated far away. Recourse to law courts was a lengthy and expensive process. It was also difficult to find witnesses to testify in their favour. The zamindars with far greater resources could easily control legal proceedings. The police and the officials of the court were in their pay. In fact, the local police officials also took their pound of flesh from the hapless ryots.

The exploitation of the ryots multiplied where the zamindari was divided into co-parcenary shares. All the claimants would then claim their dues from the ryots. In order to meet the manifold dues, the ryots had more often than not to borrow from the moneylenders who claimed more than half the principal sum as interest. Therefore, whatever was left after paying the different dues went into paying the interest. This often left the ryots with not even the bare subsistence. Coupled with this, there was also the threat that the zamindar might evict a ryot and give his land to another who promised to pay a higher rent.

The *Tattvabodhini Patrika* edited by Akshoy Kumar Dutt in the 1840s went on to highlight the anomalies in the indigo cultivation system. Unlike his mentor Rammohun, Akshoy Kumar Dutt had a clear perception of what the indigo system entailed for the cultivating ryot and the way the system operated. This was probably the first conscious presentation of the indigo system. He pointed out that the oppression of the zamindars was far surpassed by that of the indigo planters.[63] The planters gave *dadan* to the ryots and forced them to cultivate indigo either on their own land or on the land of the planters. The amount they gave was negligible and even out of this the agent and the *gomastas* took a share. The ryots were forced by the planters to cultivate indigo on their most fertile lands, lands that had until then been under rice cultivation. Even after cultivating indigo, the ryots were denied a fair price for the crop. Cultivation of indigo ruined the productivity of soil, making it unsuitable for the cultivation of other crops.

Unlike the ordinary ryots, the indigo cultivators were denied even the token means of redressal. Since the planters were mainly Europeans, the ryots could never hope to get any justice by complaining to the English magistrate. Unhindered by any legal constraints, the planters continued the forced cultivation of indigo in Bengal. The *Tattvabodhini Patrika* lamented that such a coercive system continued to persist. This description of the working and actual significance of the indigo system indicates that for the first time the Bengali intelligentsia was acquiring a clear conception of the indigo problem.

Another crucial fact, which the Tattvabodhini group brought to light, was the oppression of the rural middle class, i.e. the class that stood between the zamindars and the masses of peasantry. This 'middle class' comprised the agents and *gomastas* of the zamindar and the planters. Akshoy Kumar Dutt thus used the *Tattvabodhini Patrika*, which he edited from 1843 to 1855, to serve as a mirror to the deplorable plight, wretchedness and despondency of the peasantry of Bengal. His attacks were, in particular, directed against the native landlords and the foreign indigo planters. He wrote three essays titled 'Palligramastha Projader Durabasthya' in *Tattvabodhini Patrika* in 1850 on the miserable conditions of the Bengal peasants. Here he highlighted the oppression of the Bengal landlords and blamed the *patni* system for this.[64] In the second part of the essay, he lamented the fact that the revenue collecting officials did not take steps to protect the peasants. Here again he highlighted the oppression of the landlords.[65] In the third part he discussed at length the oppression of the indigo planters whose atrocities surpassed those of even the most powerful landlords. The indigo planters benefited from the fact that they could not be tried and brought to book by any subordinate law court existing in the *mofussil* area, because they were Europeans. Thus emboldened they terrorized, looted, beat, burnt and killed at will, if the peasants did not accept their terms. The miserable condition of the peasants made him lament:

> It is beyond anybody's conception when or how this immeasurable suffering is going to be ended. There is oppression of the landlords, tortures of the indigo-planters, exactions of the officials, and over and above, misgovernment and injustice of the government itself. Can they have any residue of their strength left, who are constant preys to such coordinated violence? They are poor in wealth, poor in knowledge, poor in religion, and even poor in strength and vigour. Is there any way to remedy this desperate and terrible situation? There exists no unity among the people; nor is there any feeling of oneness among the upper and lower classes of our society. Those who are desirous of ending the misery lack strength and power; those who have power have no will. . . . No one doubts that the Government could greatly ameliorate their present condition if they so desired. . . . But, alas, they do not move even at the sight of peasants writing under brutal tortures and oppression. Hence, they are hopelessly failing in their obligations, for which they would surely be guilty in the eyes of God.[66]

Akshoy Kumar, however, could not provide an answer to the misery of the ryots. It probably did not occur to him that the problem could not be solved without a change in the zamindari and *patni* system. He lived long enough to see further deterioration in the condition of the peasantry and did appeal to the government to redress their misery.

THE TRADITIONALISTS

If Rammohun, the other Brahmo leaders and the Tattvabodhini group constituted the Western educated moderates and the Young Bengal the Western educated radicals, then Radhakanta Deb of Shovabazar was certainly the best representative of the traditionalist group. At a time when an attack had been launched against the zamindari system, he was prompted by the need to uphold the interests of the landholding class. He was associated with the Agricultural and Horticultural Society.[67] Though he did not deal with the agrarian issues at length, he observed that the government's agrarian policy was ruinous for both the landlords and the tenants. He opposed the governments' proposal to tax the *lakhiraj* (rent-free) lands.

His contemporary, the classical humanist poet and erudite editor of *Sambad Prabhakar*, Ishwarchandra Gupta, added a new dimension to the agrarian thinking of the Bengali intelligentsia in the 1850s. Though not a product of English medium education, he imbibed much of his western liberalism from his long association with the Tattvabodhini Sabha and the eminent Brahmo leaders. Under his guidance, the *Sambad Prabhakar* became an influential paper. It adopted a very modern and progressive economic outlook. It had a close look at the agricultural scene and concluded that the government's revenue policy was responsible for the economic ruin of the zamindars and ryots of Bengal.

The journal referred to the oppression of the peasantry by the zamindars and *ijaradars* and to their subsequent ruin.[68] It stated that the greater the increase in the number of intermediaries between the zamindars and the ryots, the greater the latter's misery. The journal observed that though many held the zamindar solely responsible for the ryot's condition this was not always the case.[69] The zamindars themselves suffered greatly from the provisions of the Permanent Settlement and the sale laws. They had to pay the revenue by the sunset of a particular date. No remission was granted or any delay tolerated. Failure to comply meant the sale of the landlord's estates. Therefore, the zamindars could not afford to be forbearing to the ryots in distress. This logic in the *Prabhakar*'s argument has some validity. This attitude does not indicate a pro-landlord bias because repeatedly it highlighted in its columns the miserable condition of the ryots.

The *Sambad Prabhakar* alluded to the change[70] in the composition of the zamindari class and its effect on the zamindar-peasant relationship in the post-Permanent Settlement period. It analysed how the distress sale of lands had resulted in the entry of moneyed men from urban areas to the zamindari class. Unlike the old zamindars, they had no ties with the land and their sole aim was to make more money. This had destroyed the old relationship between zamindars and peasants. The journal[71] also alluded to the oppression of money-lenders. They advanced money and seed to the ryots at the time of cultivation and collected their dues during harvest with heavy interest. In order to pay all their dues the ryots had to continue to borrow and there was no way out of this vicious circle of debt payment for them. The only remedy was the change in the land-revenue policy of the British.

Taking its cue from the *Tattvabodhini Patrika*, the *Sambad Prabhakar* in the 1850s also systematically attacked the indigo system and the oppression of the indigo planters. It observed that the planters coerced the ryots to cultivate indigo without sufficient remuneration.[72] Their exploitation was particularly noticeable in the districts of Murshidabad, Jessore, Pabna, Dacca, Mymensingh, Bakharganj, Faridpur and Rajshahi. The oppressed indigo cultivators were denied proper justice because of the close alliance between the Magistrates and the indigo planters.[73]

Ishwarchandra's faith in British rule was, however, unshakeable. He firmly believed that all injustices committed could be removed once the British Crown was made aware of them. He wrote poems and songs addressed to Queen Victoria describing the plight of indigo ryots. He did not realize that planters and local officials were offshoots of imperialism. This was a failing he shared with many of his contemporaries to whom a Muslim alternative to British rule was not really a welcome proposition.

UPHOLDERS OF PEASANT RIGHTS

By the mid-nineteenth century under the aegis of Girish Chandra Ghosh and Harish Chandra Mookherjee, the *Hindoo Patriot* in a big way also emerged as an influential publication. Its comments on the landlord-tenant relationship are revealing:

> That relation is one from which scarcely a hundredth part of the population, of the country is free. Its extent, its influence over individual happiness and the national character, and the degree in which it modified the action of all our social and political institutions are admitted by all to be surprisingly great, yet on none of the subjects of high national interest which are presented to the Indian spectator does exist such an amount of error and prejudice.[74]

Animosity between the owners and occupiers of the soil was great and a feeling incompatible with social well-being was kept alive. The general consequence, the journal pointed out, was 'that of a fabric torn and defaced by the feuds of its inmates, where each solicitors, only for the preservation of the part inhabited by himself and intent upon the destruction of the rest'.[75] Each class tried to further its own interest at the expense of the other. The result was constant conflict in which the occupiers of land, comprising by far the large portion of the population were the greatest sufferers. However, 'it is a law of providences', the *Hindoo Patriot* observed, 'that the oppressor cannot oppress long without losing much that it is truly valuable in him and to him, and the body of the Bengal zamindars have not escaped the visitations of this law'.[76]

Therefore, it was felt that a satisfactory adjustment of the relation between the zamindar and the ryot was imperative for the good of both. These ideas expressed by the *Hindoo Patriot* found an echo in other contemporary journals. The *Hindoo Intelligencer* stated, 'Our Zamindar conceals the heart of a viper under the garb of an angel' and the 'Poorest Ryot is the first object of his rapacity'.[77] A satisfactory solution to the problem of landlord-tenant relationship was the prime necessity of the day. Discussing the cause of the miserable plight of the ryots the *Hindoo Patriot* observed:

> The principal causes to which the oppression practiced by zamindars upon their tenants may be traced are the ignorance, amounting in the case of the latter to unconsciousness of the precise nature of the change brought in their respective position by the permanent settlement and by the laws subsidiary to that measure and the abuse of the social influence of the zamindar class.[78]

The zamindar interpreted the terms of the settlement to suit personal interest. As absolute proprietor of the land he concluded that he was entitled to charge his tenants the highest rent for the use of land. The zamindar was aware that the settlement imposed some limits upon his rights, but as these were not respected or observed, it was left to the zamindar to decide on the rent within the specified time to avoid being ruined. On the other hand, the *Hindoo Patriot* observed, the Bengal tenant firmly believed that the land was the property of the person who cultivated it. He had heard that it was not the intention of the makers of the Permanent Settlement to alter either the terms of his occupancy or the terms on which he was to maintain it. He did not understand why he should ever be compelled to pay more rent for a bad harvest or any rent for a year of drought or inundation. While the tenant admitted that the zamindars had a claim on the produce of the land he could not be reconciled to the belief that his rent was liable to revision or the land he occupied to survey and measurement. However, the largest part of dispute between them originated in a misunderstanding of the exact position of the two classes. This was a very interesting point of view, which was forwarded by the *Hindoo Patriot.* It admitted the lack of clarity in the Permanent Settlement; both the classes were to blame for this.[79]

The *Hindoo Patriot*, however, observed that failure to comprehend the legal implications of the land laws would not have prompted zamindars to oppress their tenants if their had been fear of public censure on these matters.[80] So for the first time the *Hindoo Patriot* advocated the need for a strong public opinion against the oppressions of the zamindars. Not 'legal sanction' but 'social sanction', i.e. the dread of public and class opinion would alone act as a restraint on the zamindars. The oppressive landholder in Bengal lost no caste or consideration in the eyes of the public and so there was no restraint on his conduct.

On the question of the laws of summary arrest and distraint, two great evils, which the Bengali intelligentsia had condemned, the *Hindoo Patriot*, observed,[81] that on the operation of these laws depended the fate of the ryots. These laws were indispensable in the functioning of the landed economy. So for the first time there was an attempt at the justification of these laws. According to the journal, the landlords had to have their means for earning the punctual realization of their dues. However, since these two laws had been abused by the zamindars the journal advocated some changes. It stated that the ryots had already learnt the virtue of punctuality and the fear of a prolonged and expensive legal battle would ensure timely payment of their dues. The zamindars could always take recourse to law to ensure the payment of their dues. So the change in the laws of distraint and summary trial was necessary. The *Hindoo Patriot* suggested that distraint should henceforth be confined only to crops on the grounds. The tenants' home would thus be placed in comparative safety. Officers having some interest in an honest discharge of their duties should conduct the distraint procedure. This was an absolute necessity as the commissioners of distraint were thoroughly corrupt. As far as was possible the police should not be allowed to interfere. Lastly, the judicial cognizance of cases arising out of distraint should be retransferred to the civil courts with regard to changes in the law of summary arrest. The *Hindoo Patriot* suggested that the law of arrest of recalcitrant tenants should be placed upon the same footing as the general law of arrest.

The emergence of the *Hindoo Patriot* as one of the leading journals in the 1850s was due to the untiring efforts of its editor Harishchandra

Mukhopadhyay who wrote a series of essays on the agrarian problems, particularly the condition of the indigo ryots. In 1852, he became a member of the British Indian Association. As a member of this landholders' association, it was within the framework of the existing tenurial system, with the interests of the zamindars at heart that Harishchandra formulated his ideas on agrarian issues. Ever since the first publication of the *Hindoo Patriot* in 1853, he was associated with it. From 1854 until his death in 1861, he was its editor. During his tenure, the journal espoused the cause of the ryots. During the indigo revolt, lucid details were published about the oppressions of the indigo planters. He also agreed that there were cases of zamindari oppression and even talked of peasant resistance against increase of rent.[82] Yet he did not propose the abolition of the Permanent Settlement.[83] In 1861 Kristo Das Pal became the editor of *Hindoo Patriot* and held the post till 1884. During this time it actively took up the cause of the zamindars, particularly during the marathon controversy over the Tenancy Bill in the 1870s and 1880s.[84]

Another luminary of the period, Nabinkrishna Bose, the secretary of the Bethune Society became interested in the land question. He pointed out some flaws in the Permanent Settlement. It had resulted in the creation of a large number of intermediaries between the zamindars and the ryots. He pointed out the oppressive character of the zamindars and the necessity of survey of land. To ensure fair justice for all he proposed reorganization of the judiciary. He also agreed that it would be best for the ryots if the right to ownership of land were vested in them. Yet he acknowledged that this would not be possible under the present circumstances. So he advised that just as there was a settlement between the government and the zamindars, there should be a similar one between the latter and the ryots. This would, he hoped, strengthen the ryots' control over land.[85]

MUSLIMS AND CONCERN FOR THE PEASANTRY

In the meanwhile, the Farazi and Wahhabi movements swept the countryside in the 1830s. These primarily religious movements greatly impacted upon Muslim society. Though their professed aim was to purify Islam in Bengal they did include a rudimentary politico economic programme.[86]

The Wahhabi movement was a spontaneous anti-zamindar protest movement. Zamindari exploitation drove the rebels to retaliate against a Hindu zamindar at Narkelberia against whom they had an immediate quarrel. This was followed by attacks on other zamindars in the neighbourhood. Once motivated and mobilized, the rebel ryots became the sworn enemies of all local zamindars. Muslim zamindaries too were not spared. Rural moneylenders were looted and local *daroghas* were threatened. The Wahhabi movement, according to one view, presented all the features of a 'peasant-plebeian rising'.[87] The rebels also retaliated against indigo planters. Thus, the revolt was a reaction against the exploitation inherent in both the zamindari and the indigo plantation system.[88]

The Farazi movement too started as a movement for religious reform. However, it drew its main strength from the peasant masses that were drawn into the movement because it voiced some of their basic grievances and promised a radical solution to their problems. The Farazis advocated rigid adherence to Islamic tenets and strict conformity with the Koran.[89] What appealed to the peasantry was that these doctrines also contained an agrarian programme. These doctrines while distinguishing between payment of rent to a zamindar and that of tribute to the government also held that 'God made the earth common to all men, the payment of rent is contrary to his law.'[90] Therefore, they preferred landholding under the government in the *khas mahals* and sought the occupancy of the government *chars* in the large rivers.[91] The Farazis asserted a kind of equality amongst themselves that was attractive to the lower classes. They not only resisted successfully the levy of all extra or illegal cesses by zamindars and *talukdars*, they also defied collection of the land rent, often withholding it altogether. The land tax was equally hated and they began to hope that it would be abolished in future.[92]

Farazis used a religious motive to articulate their agrarian grievances. They held it illegal to pay rent to an 'infidel'. The term 'infidel' was conveniently used to apply to all Hindu zamindars and also to those Muslim zamindars who as a rule condemned Farazi practices as also the European planters.[93] While the leaders pleaded that the movement was for social reform and not really anti-government, the government that offered protection to planters and zamindars found itself in opposition.

Both movements thus articulated the interests and demands of the peasantry. The notion propounded particularly by Wahhabis that land belonged to God and the peasants were not to pay any land tax provided the rebelling peasantry with an ideology sufficient to inspire them in the fight against the oppressors. However, the movement was such as to preclude the participation of Hindu peasants, although the latter were equally the victims of zamindari oppression.

Though they launched a crusade against the zamindari system, these movements stopped short of being anything more than a revolt or reaction against the existing system. However, it was during this time that the concept of the right of peasants to ownership of land was introduced for the first time.[94] Gradually, peasants came to believe that land belonged to those who tilled it. By including an elementary agrarian programme these movements created, perhaps for the first time, an interest among the Muslim educated classes in agrarian problems, the condition of the peasantry and the question of peasant rights.

THE GENERAL ARGUMENTS

In 1859, after a long period of non-intervention, the government enacted Act X. By this time, the agrarian thinking of the Bengali intelligentsia had reached a level of maturity. They had been able to identify the major agrarian changes that had been taking place in Bengal. British policy of maximization of revenue was held responsible for the conclusion of a settlement with the zamindars. The *Hindoo Patriot* observed that, while the most rapacious zamindar waited for an occasion, a pretext to increase the rent of his lands or levy new cess upon his tenantry, the 'middle man's opportunity is perennial and his pretexts ever forthcoming'.[95] The intelligentsia emphasized the changing nature of agrarian relations. There was growing class tension between the landlords and the tenants. The intelligentsia also analysed the relationship between the peasants and the moneylenders. Commercialization had increased the cash needs of the ryots. This led to greater dependence on the moneylenders. Commercialization, however, failed to improve the lot of the peasants. The description of the working of the indigo system by the intelligentsia highlights this fact. There was increasing impoverishment of the rural masses. There was no room for capital accumulation. Consequently, there was no qualitative change in agriculture. There was no change in the technique of production. Thus, the increasing impoverishment of the masses hindered any real economic modernization.

The growing anomalies in the agrarian system had become

apparent.[96] The literati had already before them the experience of more than half a century that had elapsed since the Permanent Settlement. That the settlement was in effect a failure with regard to all its avowed objectives could be seen clearly in these years. It was supposed that the landlords would be induced to improve their farms and take greater care of cultivations, since it would mean an absolute surplus for them. However, the landlords could not unfortunately assume a character of progressive agricultural entrepreneurs. They would primarily rest content by leasing their lands out to intermediaries of various descriptions. The interests of the ryots were also not safeguarded under the Permanent Settlement. It was stipulated that the landlords would issue *pattah* and *kabuliyatnama* specifying the rates to be paid by the ryots in order to rid them of uncertainties. In most cases, the landlords did not issue these. As a result the uncertainties of the ryots continued. They were frequently subjected to new *abwab*s. The class whose interests were really protected to some extent by this settlement was the landlords as it ensured no further increase in their dues. But this relief could not be turned to the advantage of Indian agriculture.[97] This basic failure of the Permanent Settlement conspicuously escaped the notice of Rammohun and his immediate successors who were concerned with agrarian problems. Most of them were aware of the theoretical shortcomings of the settlement. Yet they supported it. They probably overlooked how horrifying could be its implementation mainly at the hands of the profiteering intermediaries. It left the ryots exposed to economic bankruptcy. Besides, they also failed to suggest concrete ways to reform agricultural practices. They did not suggest how it could be modernized. Agricultural improvement was surely one way of bettering the lot of the ryots.

NOTES

1. My capital is gone, my dwelling is in wilderness
   My wealth is gone, my self-esteem is in lost
   Now my life is in danger
   Cattle, grass, trees are lost, nothing is left
2. He was influenced by Adam Smith's *The Wealth of Nations*, Malthu's 'Essay on Population' and Ricardo's *Principles of Political Economy*. He also pursued an analysis of the *Dayabhag* and *Mitakshara* systems. The former was the Bengal school of law on inheritance while the latter was the north Indian.

3. Rammohun Roy, *Questions and Answers on the Revenue System of India*, London, 19 August 1831, p. 33.
4. Ibid.
5. Ibid.
6. Ibid.
7. Susobhan Chandra Sarkar (ed.), *Rammohun Roy on Indian Economy*, People's Publishing House, Calcutta, 1965, p. 23.
8. Roy, op. cit.
9. B.N. Ganguli, *Indian Economic Thought: Nineteenth Century Perspectives*, MacGraw Hill Publishing Co. Ltd., Delhi, 1977, pp. 47-8.
10. Sarkar (ed.), op. cit., pp. 13-14.
11. Regulation I of 1793, Sec. 8, Art. 1.
12. Ganguli, op. cit., p. 49.
13. Rammohun Roy, *Paper on the Revenue System of India*, London, 1831.
14. Ibid.
15. Ibid.
16. Ganguli, op. cit., p. 50.
17. Bhabatosh Dutta, *The Evolution of Economic Thinking in India*, Federation Hall Society, Calcutta, 1962, p. 5.
18. Rammohun Roy, 'Questions and Answers on the Revenue System of India', in *Selected Works*, op. cit.
19. Ibid.
20. Ibid.
21. Sarkar (ed.), op. cit., pp. 74-9.
22. *Sambad Kaumudi*, 26 February 1828.
23. Ibid.
24. Rammohun Roy, 'Remarks on Settlement in India by Europeans', in *Selected Works*, op. cit.
25. Ganguli, op. cit., pp. 52-5.
26. Blair B. Kling, *The Blue Mutiny: The Indigo Disturbances in Bengal, 1859-1862*, University of Pennsylvania Press, Philadelphia, 1967, p. 104.
27. *Sambad Kaumudi*, 26 February 1828.
28. H.C. Sarkar (ed.), *Life and Letters of Raja Rammohun Roy*, Classic Press, Calcutta, 1913, p. 149; S.C. Sarkar, op. cit., p. 22.
29. Suprakash Roy, *Bharater Krishak Bidroha O Ganatantrik Samgram* (Bengali), Noya Prakash, Calcutta, 1970, p. 237.
30. Sumit Sarkar, 'The Complexities of Young Bengal', *Nineteenth Century Studies*, no. 4, 1973, pp. 506-7.
31. Krishnamohan joined the missionaries. Some entered government service. Harachandra Ghosh was made *sadar amin* in Bankura in 1832, Rasik-krishna Mullick and Gobindachandra Basak became deputy-collectors around 1837-8, Chandrasekhar Deb, Sib Chandra Deb and Krishorichand

Mitra were appointed deputy-magistrates between 1843 and 1846. Peary Chand Mitra and Ramgopal Ghosh became successful businessmen. Sibnath Sastri, *Ramtanu Lahiri O Tatkalin Banga Samaj* (Bengali), op. cit., p. 125; Brojendranath, Bandyopadhyay, 'Peary Chand Mitra', *Sahitya-Sadhak Charitmala*, no. 21, Bangiya Sahitya Parishad, Calcutta, 1955, p. 189.

32. Biman Behari Majumdar, *History of Indian Social and Political Ideas from Rammohun to Dayananda*, Bookland Pvt. Ltd., Calcutta, 1967, Chap. 3; J.C Bagal, *Unobingsha Satabdir Bangla*, Bharati Library, Calcutta, 1963, pp. 188-9.
33. Manmathanath Ghosh, *Raja Dakshinaranjan Mukhopadhyay*, published by author, Calcutta, 1917, pp. 205-8.
34. Sumit Sarkar, op. cit., p. 514.
35. Peary Chand Mitra, *Ramkamal Sen*, Bengali translation edited by Jogeshchandra Bagal, Bharati Library, Calcutta, 1964, p. 60.
36. *Bengal Spectator*, April, 1842.
    Cf. From April-August 1842, this journal was published monthly. In September 1842 it became fortnightly and from March 1843 till November 1843 it was published weekly.
37. *Bengal Spectator*, 15 October 1842.
38. Ibid., 1 November; 15 November; and 15 December 1842.
39. Ibid., 17 April 1843.
40. Ibid., 8 March 1843.
41. Peary Chand Mitra, 'The Zamindar and the Ryot', *Calcutta Review*, vol. VI, no. 122.
42. Peary Chand Mitra (1818-88) was born in Calcutta. His father, Ramnarayan Mitra, was a friend of Rammohun Roy. After completing his education in Hindu College, Peary Chand was associated with the Bengal British India Society, British Indian Association, the Bengal Social Science Sabha and other welfare organizations. He was elected a member of the Agricultural and Horticultural Society of India. He achieved considerable success in internal and foreign trade as well as in the literary sphere. He wrote a series of essays in *Jnananesvana*, *Bengal Spectator*, *Calcutta Review*, *Bengal Hurkara*, *Indian Mirror*, etc. His best work was 'Alaler Gharer Dulal', *Bharatkosh*, vol. 4, Bangiya Sahitya Parishad, Calcutta, 1970, pp. 427-8.
43. *Bengal Spectator*, 1 November 1842.
44. Ibid.
45. Ibid., 25 April 1843.
46. Ibid., 16 August 1843.
47. Benoy Ghosh, *Samayikpatre Banglar Samaj Chitra*, vol. 6, Papyrus, Calcutta, 1983.
48. *Bengal Spectator*, 17 April 1843.
49. Ibid., 24 July 1843.

50. R.C. Dutt, *The Peasantry of Bengal*, ed. Narahari Kaviraj, Manisha Granthalaya Calcutta, rpt., 1980, Introduction, pp. XXX-VI
51. *Jnananesvana* articles: 'Government of the Company and Revenue System of India', rpt. in *India Gazette*, 8 April and 10 May 1833, cited in S. Sarkar, op. cit., pp. 518-19.
52. Peary Chand, Mitra, 'The Zamindar and the Ryot', *Calcutta Review*, vol. VI, July-December 1846.
53. Ibid., p. 350.
54. Ibid., pp. 351-2.
55. *Jnananesvana* extracts in *Samachar Darpan*, 21 April 1838, 26 January 1839 in Brojendranath Bandyopadhyay (ed.), *Sambadpatre Sekaler Katha*, Bangiya Sahitya Parishad, Calcutta, B.S. 1384, vol. II, pp. 331-2, 467-8.
56. Brojendranath Bandyopadhyay, 'Peary Chand Mitra', op. cit., pp. 179-80.
57. Sumit Sarkar, op. cit., p. 12.
58. Benoy Ghosh, op. cit., vol. 4, p. 43.
59. *Tattvabodhini Patrika*, Baisakh 1850.
60. Ibid.
61. Ibid., Sravana 1850.
62. Ibid.
63. Ibid., Aghrayan 1850.
64. Akshoy Kumar Dutt, 'Palligramastha Projader Durabasthya', *Tattvabodhini Patrika*, Baisakh 1772 Saka.
65. Ibid., Sravana 1772 Saka.
66. Ibid., Agrahayan 1772 Saka.
67. Jogesh Chandra Bagal, *Unobingsha Satabdir Bangla*, op. cit.
68. *Sambad Prabhakar*, 11 Ashar 1258 (B.S.).
69. Ibid., 28 Bhadra 1259 (B.S.).
70. Benoy Ghosh, op. cit., vol. I, p. 37.
71. *Sambad Prabhakar*, 5 Bhadra 1264 (B.S.).
72. Ibid, Ashar 1255 (B.S.).
73. Ibid.
74. *Hindoo Patriot*, 6 September 1855.
75. Ibid.
76. Ibid.
77. *Hindoo Intelligencer*, 18 September 1854.
78. *Hindoo Patriot*, 6 September 1855.
79. Ibid.
80. Ibid.
81. Ibid., 28 May 1857.
82. 'Lattyalism in Bengal', *Hindoo Patriot*, 9 April 1857.
83. In 1833 on the advice of George Thompson, the zamindars had founded the Land Holders Association whose aim, among others, was to uphold

the interests of the land owning class. Among its members were the Derozians Ramgopal Ghosh, Prasanna Kumar Tagore of the Pathurighata branch of the Tagore family, Raja Radhakanta Deb of Shovabazar and Joykrishna Mukherjee of Uttarpara. In 1851 it was renamed the British Indian Association. N.B. 'The Condition of the Ryot in Bengal', ibid., 16 April 1857.

84. Files of the *Hindoo Patriot*, 1861-84.
85. Nobin Kristo Bose, 'The Landed Tenure in Bengal' , in *The Proceedings of the Bethune Society*, for the session of 1859-60, 1860-61, Calcutta, 1862, pp. 45-74; Amalendu De, *Chirasthayi Bandobasta O Bangali Buddhijibi* (Bengali), Ratna Prakashan, Calcutta, 1981, pp. 16-17.
86. Amalendu De, ibid., p. 6.
87. Narahari Kaviraj, *Wahabi and Farazi Rebels of Bengal*, Peoples Publishing House, New Delhi, 1982, p. 52.
88. W.W. Hunter, *The Indian Mussalmans*, W. Rahman, Dacca, 1975, pp. 37-9.
    Though Hunter stressed that at Barasat a series of agrarian outrages took place, he does not mention that indigo planters were also not spared.
89. Judicial Progs, 29 May 1843, nos. 21-6.
90. Ibid.
91. Ibid., 25 October 1848, no. 121.
92. Report of the Commissioner of the Dacca Division, Judicial Progs, 7 April 1847, nos. 98-9.
93. Narahari Kaviraj, op. cit., pp. 92-3.
94. Amalendu De, op. cit., p. 7; Rafiuddin Ahmed, op. cit., pp. 126-30.
95. *Hindoo Patriot*, 5 March 1857.
96. Rammohun Roy, 'Revenue System of India—A Paper', op. cit.
97. Sourin Bhattacharya, 'Permanent Settlement and Rammohun Roy', *Nineteenth Century Studies*, no. 1, 1977.

CHAPTER 3

Debating the Peasant (1859-1885)

Mari durbal projar pare atyachar
Kata jane kare, kare jamindar
Tara jane mane, jamindar bine
Nahi anya keho dukkha sunibar
Praja kato sahe, kichu nahi kahe
Mane bhabe er nahi upay aar
Jamindar dhare jarimana kare
Manosadh pure, nashichey projar
Suno sabhyajan, kariye manna
Dekhaiba aji abhianay tar

MEER MASARRAF HOSSAIN[1]

After a prolonged period of non-intervention, spanning over six decades, the government enacted Act X in 1859. In view of the heightening tension between landlords and tenants the expectation that the government would soon remedy the defects of the Permanent Settlement had been high. To their newly awakened legal sense, the legal uncertainty of the peasants' position had been a matter of serious concern to the literati. There had been a growing demand for legislative interference to rectify the uncertain nature of the peasants' position. Increasing depiction in contemporary literature and journals of the exploitation of ryots and their impoverished state convinced a section of the intelligentsia of the need to force the government to realize the gravity of the situation and the need for remedial action.

THE INDIGO REBELLION

The Act X intended to remove some of the abuses of the existing law, particularly those related to distraint of crops and properties of the peasants for the realization of arrears of rent. Ryots were classified into three groups: ryots holding at fixed rates, occupancy ryots and non-occupancy ryots. Ryots holding land at a rate unchanged for twenty years before the date of rent-suits brought against them by zamindars belonged to the first group. Zamindars could not enhance their rent. Any ryot who had cultivated or held lands for a period of 12 years was an occupancy ryot, as long as he paid rent. Zamindars could enhance their rent only where cultivation had increased, or where the rent rate paid by them was lower than what was called the *pargana* rate (the prevailing rate for lands of a particular quality) or where the value of the produce had increased. These legal restraints would not apply where occupancy ryots agreed, by a written contract, to pay an enhanced rate. Ryots without the right of occupancy were not protected by the law. Zamindars could no longer compel the attendance of peasants at their courts for the adjustment of rent, or for any other purpose. Distraint was illegal for arrears of more than one years' standing or where the peasants provided security for their payment.[2] The Act thus gave a section of ryots right of occupancy and protection from arbitrary eviction provided that, they paid rents to zamindars. In this way, the government attempted to introduce relative stability among the cultivating classes and restrict grounds for rent enhancement. Rents of occupancy ryots could only be enhanced on specific grounds;[3] they could only be ejected by a judicial decree or order and their crops could only be distrained for the arrears of one year.[4] The principle of fair and equitable rent was to be the decisive factor. The court presumed that the existing rate was fair and equitable until the landlord showed the contrary. But the Act did not define what was 'fair and equitable'.[5] This caused all sorts of anomalies over rent rates as judges found it difficult to fix the zamindars' share, on the basis of this principle, in the increased value of the produce.

The Act caused a furore in zamindari circles. But the outbreak of the indigo rebellion in 1860 immediately diverted the attention of the literati. The revolt spread across the districts of Nadia, Jessore, Murshidabad, Faridpur, Malda, Pabna and Rajshahi. Sporadic and occasional expression of grievances against the unremunerative indigo cultivation had been witnessed throughout the late 1850s. The activities of Magistrate Eden of Barasat prompted the peasants to take a defiant stand in 1860. In March 1859, he upheld the peasants' right to grow whatever crops they preferred. In August 1859, he stated further that the police had a duty to protect peasants in the possession of their land even if they had failed to honour indigo contracts.[6] This inspired peasants to rebel. Supported by the small *talukdars* and the urban intelligentsia, the rebellion soon spread. But it soon suffered a temporary setback when the government revived the Regulation of 1830 by which a peasant could be sentenced for the breach of an indigo contract.[7] The immediate withdrawal of the regulation in September added momentum to the sagging movement. Having failed to force the contracted peasants to grow indigo and denied the prospect of expropriating *raiyati* land and expanding *nij* cultivation, the planters made an attempt to enhance rents in their capacity as landlords under the provisions of Act X, especially on the pretext that the value of agricultural produce had increased. The peasants retaliated with a successful no-rent campaign on the planters' estates. This had an adverse impact on zamindars and *mahajans* who had initially backed the rebels. Among the zamindars that helped were the Pal Chowdhurys of Ranaghat, Rammohun Moulik, Brindaban Sarkar and Prankristo Pal.[8] Digambar and Bishnucharan Biswas, famed leaders of the movement, came from the landed gentry. The revolt fatally affected the indigo system in Bengal. This, however, left moneylending and grain-dealing landlords as the dominant class in rural Bengal.

The struggling indigo peasants secured the sympathy of the Bengali literati. It is significant that in 1859-60 the present generation, unlike their erstwhile predecessors, viz., Rammohun or Dwarkanath Tagore no longer had any misconceptions about any apparent prosperity of indigo peasants. When the target was the European planters, their class loyalties were not questioned. The support of the landlords against the atrocities of planters helped the intelligentsia to take a definite stand on the issue. Their support came in many forms. In the columns of the pro-ryot press, detailed descriptions of the inhuman practices adopted by the planters continued to be published from the indigo districts. The *Hindoo Patriot* in 1860-1 played a leading role in this. It published lucid descriptions of the anomalous system sent anonymously by men like Girishchandra Bose, a police official posted in Krishnanagore, Manmohan Ghosh, then a student at Krishnanagore, Radhikaprasanna Mukherjee, then Deputy Inspector of Schools at Nadia and Dinabandhu Mitra, the novelist among others.[9] Sisir Kumar Ghosh, a district correspondent who toured the areas and was later to start the *Amrita Bazar Patrika*, added a new dimension to the emerging trend of investigative journalism. Harish Chandra Mookherjee, the sympathetic editor, acquired immortal fame as a champion of the rebels.[10] Appeals were made to the government to intervene. As the rebellion spread, an enquiry was sought into the state of things so that the disgraceful blemish on the administration could be immediately rectified. The refusal of the government during the tenure of J.P. Grant, the Lt.-Governor of Bengal, which was influenced by the tirade of the Christian missionaries against the inhuman practices adopted by European planters to facilitate forced cultivation, to pass legislations in favour of the planters allowed the literati the space to be vocal in their denouncement of the indigo system.[11] Prompted by the Christian missionaries they vent their ire through the native press, against a system that hurt their sensibility.[12] The revolt gave them an opportunity to emerge as champions of the social downtrodden. The fact that their loyalty to the Raj was not questioned eased their situation and was very much in keeping with their stand since the Revolt of 1857.

In powerful prose, in a language that was significantly refined, Dinabandhu Mitra, the author of *Nil Darpan* published in 1860, a literary creation that was synonymous with the revolt, denounced the oppression of peasants.[13] As Inspecting Post Master, the author had first-hand knowledge of the prevailing situation in the indigo districts. This gave him a deep insight into the hapless lives of the ryots. The powerful resistance put up by Torap made a gripping tale. This reduced the gap between the educated urban intelligentsia and the rural poor and took the story of the latter's plight from the country fields to the well-decorated parlours. The fact that the author was able to convince Reverend James Long to depose in his favour added credibility to such efforts.

The literati recognized the determination and courage displayed by the rebelling ryots in the face of stiff opposition by the planters. Soon help poured into the countryside in the form of legal and financial aid. Harish Chandra observed with approval in 1860 that though lacking in power, wealth and political knowledge, the Bengal peasantry had affected a revolution, which was magnificent in dimension, and significant in character.[14] It was due to his efforts that the British Indian Association took up the cause of the peasants and appointed legal aides to plead their cases in courts. Harish Chandra himself worked tirelessly rendering judicious advice, drafting petitions and letters and overseeing their cases in the law courts. His home was open to the ryots who came to seek his help and advice.[15] Sisir Kumar Ghosh proved to be an equal partner in this. His feelings for the struggling ryots, insight into their pain, were sharpened by his own experience. His own father, a lawyer, had been victimized for successfully defending a ryot against the oppression of a European planter. He toured from village to village trying to unite the ryots and make them realize the anomalies of the indigo system. He gave them legal advice and was recognized by the ryots as their saviour.[16]

A section of the rural educated society was soon actively engaged in inciting peasants and keeping them united during the rebellion in the face of a spirited attempt by planters to scare the latter. Physical protection and financial support was extended. Help came from other quarters also. Young intellectuals employed in government service came forward to render support. Notable among them were Maulavi Abdul Latif, deputy magistrate of Kalaraoh,[17] Chandramohan Chatterjee, the deputy magistrate of Murshidabad and Bankimchandra Chattopadhyay, Bengal's foremost litterateur, who was then the magistrate of Khulna. They gave administrative support and did yeomen service in checking the oppression of planters.

Such overwhelming support helped the cause of the struggling peasants. The government was forced to take notice and appoint an Indigo Commission. Government policy based on its findings helped the cultivation of indigo in Bengal to die a natural death.

ACT X AND ITS AFTERMATH

Once the indigo districts quietened down, attention was diverted to the clauses of Act X. In the period that followed, landlords carried on a prolonged crusade against it. Under the new law, enhancement of rent by a zamindar was very difficult since he had to furnish the impossible proof that the rent enhancement that was demanded by him was in the same proportion in which the increase in the value of produce had taken place. In fact, the landlords thus found it difficult to enforce the right of enhancement while the ryots were saddled with the right of occupancy which they could not prove. The Act thus came under fire from all quarters.

Joykrishna Mukherjee of Uttarpara, the articulate spokesperson of landlords, denounced the Act as 'an apple of discord'. It had vitiated the entire relationship between them and the ryots and asked for its immediate revision.[18] The British Indian Association, a body that claimed to be the representative of Indians ever since its inception but was in reality the custodian of landholding interests, complained in the 1860s that the existing Act was not precise, distinct or adequate. As a result, neither the tenants nor the landlords could clearly comprehend their respective rights.[19] Consequently, the secretary of the Association demanded that the law be suitably amended, greater facilities be given for the proper realization of rents and definite rules for the enhancement of rents be laid down.

The main argument of the landlords was that it constituted a direct violation of the Permanent Settlement. Joy Krishna Mukherjee held that some sections of the Act infringed on the zamindar's rights and privileges and reduced him to the position of the mere tax collector.[20] At a meeting of the British Indian Association on 28 July 1865, he stated that it had benefited neither zamindars nor ryots. It had only succeeded in introducing great confusion and uncertainty into a system of land tenure that caused a split in zamindar-tenant relationship forever.

Landlords resented in particular sections VI and XI which declared that they could no longer compel attendance of peasants at their courts for the adjustment of rent or for any other purpose. They also resented the creation of occupancy rights in land. The East Bengal Landholders' Association complained that while bestowing the right of occupancy on ryots it deprived zamindars of a right conferred by the Permanent Settlement.[21] Landholders of Nadia sent a petition to the legislative council. It asked for amendment of those sections of the Act that were detrimental to their interests.[22] Resentment of the landlords was due to the fact that the right of occupancy was closely linked with the right of enhancement which they thought was greatly limited by Act X. They desired immediate withdrawal of the former.[23]

To the ryots, Act X was not an unmixed blessing. Under it, the new occupancy ryot was required to furnish the almost impossible proof that he had held every particular field of his holding for twelve years.[24] The zamindars moreover, could easily evade the law by shifting his tenants from one field to another in order to prevent the accrual of occupancy rights. Consequently, capricious zamindars often successfully forced ryots to give up their occupancy title. The Act also, neither gave occupancy ryots any protection from incessant enhancements, nor determined any rules for fixing an equitable rate of rent. The law did not define the occupancy ryots' right to improvements or determine his right in them in the event of his eviction.

As for the ryots, they had no platform of their own from which to voice their grievances and demands. They had no association of their own. The Act did not satisfy them. The large sections of under-ryots were ignored by it. However, they were determined to thwart all attempts of the landholders to have the right of occupancy repealed. The mounting tension between them would soon draw attention to the countryside.

The British civilians desired a change because the 'agrarian conflicts not only disturbed the social equilibrium in the districts but also caused great administrative difficulties'.[25] The intelligentsia too was drawn into this bitter controversy affecting landlord-tenant relationship. Both classes found their supporters among them. Those who criticized the Act were frankly partisan in their attitudes. Many of them strived to protect their vested interests by expressing solidarity with the landholding classes. However, there was also a section that emerged as champions of the interests of ryots. In the 1860s and 1870s, they struggled to safeguard the interests of the latter and were vocal in their condemnation of the exploitation of the zamindars. Depending on whose interests they supported, the intelligentsia came to be broadly distinguished as pro-landlord and pro-ryot.

Their handling of the problem of land-tenant relationship is to be studied in two uneven phases. The first phases constitutes the twenty-year period following the enactment of Act X. In the second phase is to be seen how the two sections tried to safeguard the interests of the agrarian class they were defending during the marathon rent controversy between 1880-5, inaugurated by the appointment of the Rent Law Commission which culminated in the enactment of Bengal Tenancy Act in 1885. This was a chance to ensure the practical application of their respective claims.

INTELLIGENTSIA SUPPORT FOR LANDLORDS

Intelligentsia support for landlords was rendered through their associations, representations in legislative bodies and the press. It was the British Indian Association that first came out in support of landholding interests. Its membership register recorded the names of the erstwhile landlord and well-known educationist Radhakanta Deb, the Derozian Ramgopal Ghosh, landlord and lawyer Prasanna Kumar Tagore and the Calcutta High Court Judge Sambhunath Pandit. Its landlord members were determined in the 1860s and

1870s to keep intact every clause of the Permanent Settlement. It was based on an alliance between landlords and merchants and a segment of the intellectuals. An alliance between landlords and merchants was easily achieved since, they belonged to the same class and; at many points the two sections overlapped. It was customary for prosperous merchants to invest in land. But the intellectuals who became members were not landowners in their own right. By upholding the cause of landholders, most of who lived in luxury and indolence, they lost the right and power to lead the rapidly growing middle class in the last quarter of the century.

Kristo Das Pal is a case in point. In 1858 he was appointed Assistant Secretary of the British Indian Association. Very early in his life, he made a mark as journalist and after the death of Harish Chandra became editor of *Hindoo Patriot* in 1861. From then till Pal's death in 1884 the *Patriot* was to act as a champion of the rights of landlords. Wielding great influence on zamindars, in his dual capacity he found ample outlet for his skill and for almost a decade he enjoyed political leadership. In 1872, the government nominated him as a member of the Bengal Legislative Council. When the Bengal Tenancy Bill was considered the Governor-General Lord Ripon entrusted the British Indian Association of appointing a representative of zamindars to his Council. K.D. Pal was chosen for the job. Week after week he conducted an agitation to prevent infringements on the Permanent Settlement. The civil courts were blamed for seriously encroaching upon the rights of landlords even while interpreting the new Act for dispensing justice.[26] The *Hindoo Patriot* thus concluded: 'The old good feeling between the zamindar and ryot had been destroyed; the latter who used to look up to the former as his *mabap*, no longer cherishes him the same regard; discord had taken the place of harmony and hostility that of amity.'[27]

Sisir Kumar Ghosh, in stark contrast to his earlier stand on the peasant, endorsed this view. His *Amrita Bazar Patrika* and the other pro-landlord publication *Spectator* accused the government of depriving zamindars of many of their old privileges, viz., the right to compel tenants to attend the estate office and facilitating ryots' recourse to the law courts on the flimsiest of grounds.[28]

The resentment of landlords sharpened during the agrarian unrest that started in the Yusufzahi Pargana in Pabna district in 1873 and soon spread like bushfire in other parts of central and eastern Bengal. The main issues of conflict became occupancy rights and rent. Some rentier landlords, many of whom controlled substantial demesne and were also engaged in profitable moneylending and grain-dealing, sought to invalidate the ryot friendly clauses of Act X and increase rent in an era of rising prices. Throughout the 1860s and early 1870s they also attempted to prevent the accrual of occupancy rights that would have entailed drastic moderation of rents, by shifting their tenants around different holdings in the village and forcing them to accept new *kabuliyats* thereby foregoing their legal rights to occupancy status. As the new Act imposed restrictions on enhancement of rent, they tried to incorporate the various *abwabs* in the legal rent in the new *kabuliyats*. A conscious tenantry was determined to combine and resist all these manipulations. In May 1873, an agrarian league was formed in the Yusufzahi Pargana of the Pabna district to resist the impositions of the Banerjee, Tagore and Sanyal zamindari families of the area.[29] Refusing to pay rent on one hand and filing petitions to seek redress against illegal zamindari exactions, soon the peasants in Bogra, Rajshahi, Dhaka, Mymensingh, Faridpur, Tipperah and Bakharganj followed their counterparts in Serajgunge.[30]

As the movement gathered momentum, the pro-landlord press held the government responsible for the situation by instilling in the rebels the determination to fight for their rights after years of docility.[31] *Desh Hitoishini* suggested a scheme of reforms:

> Let the land be divided into different classes according to its productive quality and a certain rate of rent be fixed on each class. The zamindars will be bound to levy only those rents and the ryots to pay them under the orders of government. Neither party will be distressed nor the zamindars will make themselves liable to punishment if they try to extort more. . . . If some such arrangement is not made and the Permanent Settlement rights infringed upon, neither party will be benefited; on the contrary it will tend to increase the existing ill-feeling.[32]

Hindoo Ranjika, edited by Srinath Singha, the president of Boalia Dharmasabha of Dacca, pleaded for the revision of Act X since it had rendered great injustice to zamindars.[33] This was in fact the overall demand of the intelligentsia promoting the interests of the gentry. They wanted pre-1859 *status quo* to be maintained. They demanded for the landlords unrestrained right to rent enhancement, eviction and distraint and dismissed the need for any form of occupancy rights for peasants.

In the meanwhile, zamindars realized by the 1870s that mere coercive measures were not sufficient to frighten ryots to submission.

They were also sceptical about the sincerity of government intention to quell the rebellion. Poet and elder son of Maharshi Debendranath Tagore, Dwijendranath who for a brief spell was in-charge of the Tagore estates, in a letter to the government complained that the zamindars were in fear of their life and property and sought government intervention in quelling the ryots.[34]

On their own a scared gentry sought to confront the unprecedented situation by framing since 1874 a large number of perpetual leases. Faced with growing difficulties in rent collection they decided to share agricultural profits with the under-tenure holders instead of bearing all the trouble for collection of the same. Therefore, they pressurized the government to compel occupancy ryots to part with them a portion of the profit. In 1876, in course of the debates on the Agrarian Disputes Act, Pal urged that indefiniteness of Act X had brought rent suits to a deadlock.[35] Lt.-Governor Temple agreed to the need for formulating some principles for determination of rent that was fixed by legislation. In August 1876 he proposed to introduce a Bill to define the principle on which the right of occupancy ryots and tenure-holders should be fixed, to simplify the procedure for realizing arrears of rent in undisputed cases, to extend definition of occupancy ryots and to strengthen their rights in land.[36] Temple suggested that the rent of occupancy ryots be fixed at rates less by at least 25 per cent than the rate ordinarily paid by non-occupancy ryots. In case of difficulty, such rent was to be calculated at 20 per cent of the value of gross produce as the basis for determining the rent of an occupancy ryot. This would mean that an occupancy ryots' rent, calculated on that basis and being at least 25 per cent less would be 15 per cent of the value of the gross produce.[37]

The proposals of Temple were not, however, fully considered when early in 1877; Sir Ashley Eden (1877-82) took over as Lt.-Governor. It was then arranged that larger amendment of the law should be deferred, and a bill providing only for the realization of undisputed arrears be introduced at once.[38] Eden was anxious to formulate a summary procedure to enable them to overcome the passive resistance of their ryots.[39] But he also desired that ryots should enjoy those rights that the ancient land law and custom of the country intended them to have, protected against arbitrary eviction and left with a share of the profits of their toil.[40] But the government found it impossible to frame a procedure which was acceptable to all.[41]

It was, however, found when the bill for the recovery of arrears of rent was introduced, that it was impracticable to limit its scope. The Bill for speedy realization of arrear rents evoked widespread protest. It failed to satisfy either ryots or zamindars. The latter were dissatisfied with provisions relating to occupancy tenures. Their contention was that the right of transferability would not be conducive to the interests of either class. The Bill was then virtually dropped and the Lt.-Governor appointed a Rent Law Commission in 1879 to review the tenurial system in its entirety.

THE PRO-RYOT PRESS

A section of the intelligentsia in this controversy adopted a liberal stand and widely used the press to express them. Mostly engaged in the learned professions and government service they were no less articulate. It was with deep regret that the landed gentry noticed that, 'unfortunately a portion of our educated young countrymen seem to sympathise with the ryots'.[42] However, while they debated and explained the ryots' cause they failed to get to the depth of the agrarian problem and offer a radical solution. They refuted the charge that Act X constituted an encroachment into the rights of landlords. Both *Sadharani* and *Som Prakash* defended it on the ground that Section 8 of Act I of 1793 that was prompted by realization of the helplessness of peasants had clearly enunciated the intentions of the framers. Both Act X of 1859 and Act VIII of 1869 had been enacted in pursuance of this.[43] The *Bengalee* defended the former explaining that it was 'forced on the government of Lord Canning by a mass of evidences showing that the old regulations had worked oppressively'.[44] The *Bengal Magazine* echoed the same.[45] Denouncing the attempts to invalidate the Act the *Reis and Rayyet* thundered that 'whatever grasping landlords may say the peasantry of a country have as good a claim on the protection of government as they. Nor can it be denied that, if they had done their duty fairly by the tenantry, the unpleasantness of state intervention between them in permanently settled districts would not have arisen'.[46] The *Sulabh Samachar* asked for a redefinition of the respective rights of ryots.[47] *Sahachar* adopted a more radical stand. Asserting that zamindars could not be recognized as sole proprietors of land, it wanted a ceiling be fixed on agricultural holdings and asked that the increased value of the land be appropriated not just by zamindars but be shared by ryots.[48] What is crucial here is that zamindars and ryots were perceived as co-partners in the agricultural process.

As to the leadership and control of the journals defending the ryots' cause, most of them belonged to the professional classes. Akshaychandra Sarkar (1846-1917), the founder-editor of the weekly *Sadharani*, was educated at Presidency College. A lawyer by profession, he was founder-member and assistant secretary of the Indian Association.[49] Krishnamohan Banerjee (1813-85), founder-editor of *Enquirer*, was founder-member of Academic Association and Society for the Acquisition of General Knowledge and also member and President of the Indian Association. Rangalal Banerjee (1827-87), also associated with the Indian Association, edited *Education Gazette*. An eminent poet he joined government service after completing his education at Mohsin College, Chinsurah. Surendranath Banerjee (1848-1925) was educated at Doveton College, Calcutta and Calcutta University. He joined the Indian Civil Service but was dismissed from service. A teacher at Metropolitan and City College, he was a founder-member and secretary of the Indian Association and proprietor and editor of *Bengalee*. Krishnakumar Mitra (1852-1937), the editor of *Sanjivani* was a member and assistant secretary of the Indian Association. He was educated at Mymensingh District School and Presidency College and was a teacher at City College. *Reis and Rayyet* was edited by Sambhuchandra Mukherjee (1839-94). He also edited *Mookherjee's Magazine* and was a teacher by profession. Associated with the *Morning Chronicle* and *Samachar Hindoostani*, he was an author of considerable repute.[50] Dwarkanath Vidyabhushan (1820-86), educated at Sanskrit College, was teacher and founder-editor of *Som Prakash*. He was a teacher of Sanskrit College who edited *Som Prakash* from its inception in 1862 till his death in August 1882 except for a brief spell when it was edited by Mohanlal Vidyabagish and Dwarkanath's nephew Sibnath Sastri.

These Bengali liberals, however, could not be unanimous on the benefits accrued to ryots by the Rent Act. While Akshaychandra Sarkar of *Sadharani* hoped it would serve as a 'fort for the protection of ryots'[51] and Surendranath Banerjee regarded it as the Magna Carta of the cultivating classes,[52] there were others who found serious flaws in it.[53] Dwarkanath Vidyabhushan persistently pleaded for a revision of the tenant law and for the enactment of a permanent settlement between landlords and tenants. He observed that agricultural production had not increased, agrarian activities had deteriorated and the sufferings of the ryots were manifold.[54] The *Bengal Magazine* in support of his view observed that zamindars, seeing their rights clearly defined in it had resorted to enhancement whenever convenient and owing to their superior knowledge and wealth had succeeded in many instances in obtaining all they wanted.[55] The *Sambad Prabhakar* condemned the clauses in Act X that allowed enhancement of rent on the ground of increase in the value of the produce.[56] This policy of 'Political Economy' followed by the government was the source of immense suffering to the ryots wrote *Sadharani*.[57]

Unexpected support came from *The Pioneer*. This Allahabad newspaper, which generally supported official policies summed up the actual impact of the Act and wrote: 'But with the law now as it stands, it is pretty clear that the zamindars will have the best; the ryots the worst of a pitched battle. The Act X without knowing is a Magna Carta for the zamindars. This Act can be used against the ryots as an engine of oppression in the matter of enhancement of rent.'[58]

The entire Act fell far short of the expectations of most members of the intelligentsia who concluded that it had failed to provide a satisfactory definition of landlord-tenant relationship. In spite of creating occupancy right in land it failed to evolve a proper rule for enhancement of rent.[59] They were particularly critical of the clause that allowed ryots to contract themselves out of the occupancy rights.[60] Attributing all agrarian struggles from the Rungpore rebellion of 1783 to that in Pabna in 1873 to enhancement of rent in an article titled 'What Kind of Settlement Ought to be Made Between the Zamindar and the Ryot?', the *Som Prakash* asked for a Permanent Settlement between landlords and tenants.[61] The expenses and profits of ryots, zamindars and the government were to be considered to eliminate all causes for dissatisfaction. It asked for legal provisions to prevent zamindars from arbitrarily enhancing rents and ryots from refusing 'to pay their just dues'.[62] This would put adequate checks on the continual increase in the number of rent cases.[63]

Perhaps the most vocal crusader of interests of ryots in this phase of peasant struggle for justice in 1860s and 1870s was Harinath Majumdar, the articulate editor of the immensely popular *Grambarta Prakashika* appropriately named because of its avowed intention of publishing news from the countryside. Harinath (1833-96) was a self-taught man who was denied a formal education because of abject poverty.[64] However, life in the countryside and experiences gained while working in indigo factories gave him valuable insight into the rural world and lives of the peasants. He alone among all his contemporaries, particularly those engaged in journalistic pursuits, lived his life in the countryside and had a direct contact with the world of the peasantry. In the neighbourhood of his native village of Kumarkhali, many big zamindars had their estates. Notable among them were the Bhaduris, Tagores, Sanyals and Banerjees. Witness to the exploitative character of the zamindari system he published horrid tales of the activities of zamindars and their estate officials.[65] He blamed the faulty legal system and the nexus between zamindars, police, and local administrative officials for adding to the woes of the peasants. Peasants were subject to physical torture, looting of houses and farm animals and confiscation of grains. Difficulties of finding witnesses and inactivity of the local police made legal redress difficult.[66] Though he conceded that there were both good and bad landowners, everywhere, most zamindars were merciless in collecting their dues.[67] The latter far outnumbered the former. Harinath enumerated in great details the exactions imposed on peasants. He was critical of the system that had facilitated the emergence of innumerable layers of intermediaries. Till the outbreak of agrarian unrest in 1873 he said it was the docility and 'peace-loving' nature of Bengal peasantry that had prompted them to tolerate such exploitation.[68] *Grambarta Prakashika* had such a powerful impact on the peasant psyche that Harinath was inundated with innumerable pleas by the suffering peasants to write about their woes.[69]

His effort to depict zamindari exploitation brought Harinath into conflict with local zamindars. It is reported that they even sent *lathials* to force him stop publication of his reports. The local ryots created a protective shield round him and the zamindars and their men could not frighten him into submission.[70] During the agrarian disturbances in 1873 he whole-heartedly supported the rebelling peasants and was severely critical of the stand taken by K.D. Pal and Sisir Kumar Ghosh. He blamed the revolt on excessive increases in rental rates.[71] As remedial action he suggested the appointment of a Commission, consisting of neutral and knowledgeable persons to represent the peasants, to enquire into the causes of dispute as the only way to help the warring factions. He advised the government to fix the yardstick for the measurement of land and the rate of rent to stop agrarian relations from being further vitiated.[72] He advised modification of Act X for in his perception it denied minimum rights to the vast majority of peasants. They bore the brunt of the loss in the event of bad harvests. Besides, as zamindars were the immediate oppressors of tenants, permanent checks on the powers of the former were necessary. Harinath, unlike others, was also aware of the need for basic education for ryots, to make them aware of their rights and convince them to look after their own interests. He advocated that their relationship with zamindars be properly defined and regulated by law and they be provided with easy means of redressal of their grievances. He felt it imperative for the government to play an active role in this and inspire zamindars to discharge the paternalistic role envisaged for them by the framers.[73] What is significant is that even while denouncing the exploitation of peasants he distinguished between good and bad zamindars.

In fact, while the liberal intelligentsia wanted a new rent law to protect the minimum rights of tenants guaranteed under the existing law they wanted no basic and fundamental change in property relations. They wanted greater security for tenants within the existing framework of the land settlement.[74] As *Soṃ Prakash* elucidated: 'We do not want to see one class benefited at the cost of another. On the contrary, it is our prayer that impartial justice may be meted out—that the weak be protected from oppressions, that the night of trouble of the ignorant and dumb ryots may forever pass away.'[75]

Other vernacular newspapers expressed similar views[76] but there were some whose perception of the existing scenario was rather confused. The *Rajshahi Samachar* stated: 'We are neither of those that seek to recover for the zamindars the rights lost to them by the passing of Act X, nor of those that proposed to obtain an extension of the rights of the tenantry.'[77]

This would surmise to be an acceptance of the changes made in 1859 without seeking further modification. Nevertheless, these were lone voices. The general trend was an expression of dissatisfaction and demand for further modifications even while reluctant to condone any form of violent activities on the part of ryots or threaten established property structure. Even the *Bengalee* that consistently espoused the cause of the tenantry wrote during the height of the Pabna revolt condemning the display of violent anger:

> Though our sympathies are invariably in favour of the oppressed against the oppressor, we have none to spare for the Pabna ryots. If they were oppressed by their zamindars they had the courts of justice open to them. Nothing can palliate, much less excuse the outrages which are committed by them.[78]

This indicates the readiness to lead a particular form of struggle even while making the best use of the opportunity provided by the outbreak of agrarian unrest to emerge as spokesman of the multitudes of oppressed peasantry and seeking to find a foothold in the country-side as the nationalist movement began to gain ground.

In the meanwhile, relations between zamindars and tenant ryots in east and north Bengal had continued to deteriorate. Lt.-Governor Temple (1874-7) was convinced of the need to find a satisfactory solution to the rent question. On 10 July 1876, the Agrarian Disputes Act V was passed[79] in the face of stiff resistance from contemporary journals. They feared that the new Act would ruin the rights accrued to tenant ryots by the Laws of 1859 and 1869.[80] The Indian Association protested that transference of enhancement suits from civil courts, as prescribed by the new law, to the already overburdened Collectors would not be conducive to strict and proper justice.[81]

The pro-ryot section of the vernacular press launched a vehement critique of Temple's proposals. The *Dacca Prakash*[82] accused him of having adopted an anti-ryot stand. Surendranath Banerjee called him a feeble administrator induced by zamindars to ignore the solemn pledge to ryots contained in Act X.[83] Vidyabhushan feared that the proposed measure would only ruin the tenantry by offering greater facilities and power to zamindars.[84] The pro-zamindar *Amrita Bazar Patrika* expressed its dissatisfaction with the Act in even harsher terms by warning the government of an impending rebellion in the Province.[85] In fact, the vernacular press was mostly in favour of the withdrawal of the Act.[86]

Temple's successor Eden's attempt to introduce a bill for the recovery of the rental arrears further infuriated the pro-ryot intelligentsia. They apprehended that land would now easily be transferred to the moneylending section. The *Bengalee* exulted: 'We are threatened with all sorts of evil if the bill is passed; ryots will be reduced to homeless beggary; lands will pass into the hands of the moneylender; perhaps the arrangement of the solar system will be destroyed if occupancy rights are made transferable.'[87]

Ryots contended that the summary procedure laid down in the bill would virtually deprive them of the right to defend rent suits instituted against them by landlords.[88] Ultimately when the bill was dropped and a Rent Law Commission appointed in 1879 there was extreme satisfaction among the ryots and their sympathizers.[89] The fear that the pro-ryot sections of Act X would be invalidated gave way to a new hope of a better deal for the hapless tenantry.

LITERARY REPRESENTATION OF THE PEASANT AND THE RURAL WORLD

This period also witnessed an outpouring of creative literature in Bengal. Their authors came from diverse social backgrounds. Some of them were eminent intellectuals of their time whose works reflected their conscious thinking on issues that they prioritized. In most cases their portrayal of the peasantry was a manifestation of their coherent agrarian thinking. The interest centred on the legal and social status of peasants, types of tenures and their obligations, organization of agrarian production, impact of world economic forces on the agrarian economy and existing land legislations. These were men who were highly educated and professionals. For their depiction of rural life they adopted diverse forms of literary genres. These works focused on increasing rural impoverishment and argued for improvement in material conditions of the peasant within the existing framework of the land system. The contextualization of these texts in their social milieu is crucial for an understanding of the purpose of their writers.

The peasant and rural society gradually came to occupy centre-stage in different literary genres from the play to the fiction. The ordinary men of the countryside became the central characters. There was an attempt to portray the joys, sorrows, rigours of daily lives and the aspirations of the peasants. In the poetry that was produced, illustration of the writers' concern in particular for the poverty of the peasant is apparent. Men who came from a rural background or were educated in the vernacular medium also produced literature dealing with similar themes. There were also the semi-literates whose writings reflect an unconscious outpouring of their anguish at the actual realities of rural life as they perceived it. The innumerable tract literature that is available bears ample testimony to this. The abundance and gratuitous nature of some of these sources add to their authenticity.[90] Representation of the peasant in these texts differed from that in the more sophisticated literature in the issues that they tackled.

With the publication of Dinabandhu Mitra's *Nil Darpan* peasant resistance to a grossly unfair system became a theme for literary expression in this period. Sanjibchandra Chattopadhyay introduced a new style in the history of Bengali literature in 1863-4 when he used the prose medium to discuss the rights and liabilities of the Bengal ryots. His *Bengal Ryots: Their Rights and Liabilities: Being an Elementary Treatise on the Law of Landlord and Tenant* was published in 1864.[91] The elder brother of the more famous Bankimchandra, Sanjibchandra (1834-93) rose to the rank of a Deputy Magistrate and was also a talented writer in his own right who regularly wrote in *Bangadarshan*.[92]

Inspired by the ideas of Mill, Bentham and Malthus, in his examination of the law of landlord and tenant, Sanjibchandra aspired to examine 'the principles which have guided legislation' on the issue of landlord and tenant in Bengal, the procedural details and the historic changes that in the process of time revolutionized the legal and social relations between agrarian classes. He focused on the condition of ryots through the ages. While admitting that he did not have much information on the 'Hindu period' he asserted that their situation was 'wretched' during 'Muslim rule'. There was no significant improvement until 1859. He was critical of the Permanent Settlement for giving zamindars proprietary powers on land, right to rent enhancement and eviction. In his understanding of the pre-British period, in spite of rigorous exploitation, ryots had their rights of occupancy and that of holding land at the established rent. But the Permanent Settlement which overlooked them ensured the steady deterioration in their material condition by giving to zamindars rights that were abused 'as a license for unlimited exactions and oppressions'.

Sanjibchandra accepted the basic premise of the Settlement that the claims of the government on the produce of the soil should be permanently fixed and the proprietorship be vested in private individuals interested in improvement of land. But the Settlement did not give the right to those who were most likely to prove the greatest improvers. The fundamental right and obligation of the ryot, he observed, were simple. The tenant who used the land of another for his own benefit, he said, had a single right and a single obligation. His obligation was to pay rent to him whose land he enjoyed. Once the rent was paid, he had the right to enjoy the remainder of the fruits of his labour. To enjoy this, he should be allowed to retain the land he cultivated unhindered by the whims of landlords. He defended the right of occupancy for ryots against the power of the rentier class to deprive him of land.[93] He wrote a critique of the Regulations of 1793, 1799 and 1812 which gave the final shape to the Permanent Settlement and welcomed the changes made by Act X of 1859.[94] He was particularly appreciative of those provisions that checked arbitrary exactions of zamindars, insisted on exchange of *pattahs* and *kabuliyats* and restrictions of rent enhancements and eviction.[95] He discussed all the clauses of the Act from the standpoint of a jurist. The rightless condition of tenants-at-will and non-occupancy ryots did not escape his attention. He admitted that whatever benefits were extended to ryots by Act X accrued only to occupancy ryots. He was also critical of the fact that no legislative provisions were made defining the tenants' right to sale. The matter was determined by usage. The Act merely provided for registration of transfers in the zamindars' *sheristah*. Its only object was that the zamindar might have information as to who his tenant was.[96]

Sanjibchandra's work dealt with 'rights' and 'liabilities' of ryots at a time when his contemporaries dealing with the history of the revenue system in Bengal generally approached the subject from the standpoint of government-zamindar relations and confined themselves to incidental references so far as the impact on ryots was concerned. This was the best exposition on the rights of ryots. The comprehensive method he adopted in describing the legal and social relation between zamindars and ryots speaks of his profound concern for the latter. He studied the 'antecedents' and 'precursors' of the Permanent Settlement and illustrated its working by an analysis of its legislative history. However, he failed to understand the emergence of a new class belonging to the middle strata that was dependent on the new land system. He remained silent on the need to abolish this Settlement and effect changes in intermediary tenures. Like many of his contemporaries, he too was handicapped by misconception about the true implications of colonial rule. The image of a benevolent and well-intentioned administration was too deeply ingrained to be easily shaken. He probably hoped to ensure moderation and mitigation of the severity of landlords' demands and strengthen the rights of ryots by recapitulating the grounds upon which ryots' rights were based. However, none of his later works referred to agrarian issues. He was strangely silent during the marathon rent controversy in the 1880s preceding the Tenancy Act.

The best literary representation of the peasant in nineteenth-century prose was 'Bangadesher Krishak' by his younger brother Bankimchandra in 1872 in *Bangadarshan*.[97] Using several literary genres, the great novelist and satirist ushered in the emergence of the novel genre with formal realism. Acquaintance with Western social philosophy and a firsthand experience of rural Bengal equipped him to raise certain pertinent social and economic issues. He was a product of Hindu College and Calcutta University. It opened the doors of government service. He joined as deputy collector and rose to the rank of district magistrate. This gave him the opportunity to travel and work in thirteen districts of Bengal most of which are in present West Bengal, Orissa as also Jessore and Khulna, the western districts of Bangladesh. His acquaintance with the exploited Muslim peasantry was nominal. This probably explains why his fictitious oppressed peasant was a Paran Mandal.[98] This was in spite of the fact that most of Bengal's' impoverished peasants were Muslims. The content of this essay was a marked departure from the subjects he had hitherto dealt with in his earlier writings. Till then he had focused primarily on social and historical subjects. But with 'Bangadesher Krishak' the peasant henceforth became the subject of literary discourse.

In the essay, Bankimchandra examined some of the hotly debated agrarian issues of the day, zamindari exploitation and the problems and grievances of the peasantry. He begins with a description of how British rule had benefited the country. It had provided peace and prosperity. Increasing agricultural production had ensured an increase in agricultural and national wealth. But it was the state, the zamindars, moneylenders and traders who shared this wealth. The uneducated peasantry was represented as easy victims of the oppression of the 'intellectually superior' classes. His central argument was that the Permanent Settlement should have been made not with the class of unproductive zamindars but with the tenants.[99] It was a land settlement that could only have worked if zamindars were kind and sympathetic to their tenants but this was an unrealistic expectation. What had happened was that greed and rapacity of a certain section of landlords,[100] instances of which he catalogued at great length, led to their impoverishment.

The most moving passages in this essay are those dealing with the condition of the peasantry. He wrote:

> The goddess of agriculture is smiling on the land. Wealth is raining down by her grace. . . . Everyone prosper through that prosperity—the king, landlords, merchants, moneylenders, everyone, only there is no prosperity for the peasant. The bigger animals like the Tiger devour smaller animals like goats. . . . The big man known as the zamindar devours the small man called the peasant.[101]

He defended his stringent criticism of Bengal zamindars thus: 'Let that voice which fails to cry out in pain for those suffering pain be choked. Let the pen which fails to write for the benefit of people (living) in misery dry up.'[102]

He claimed that his portrait of Paran Mandal—a description of unending harassment by both big and small exploiters and consequent hopelessness—was true to life.

The focus is on unending poverty. Increase in population caused an increase in the demand for land thereby forcing up the rent. After meeting the dues of zamindars and moneylenders, ryots were left with fairly little. The novelist drew up a list of different type of cesses extracted by zamindars, estate officials, priests, guards and sundry others. Apart from these regular dues, ryots were also bound to bear the expenses of taxes imposed by the government on zamindars, viz., road and hospital cesses, income tax, etc. Bankimchandra realized that the creation and legal recognition accorded to intermediary tenures caused greater difficulties.

His heart-rending account of what the zamindari system had done to the mass of the peasantry concludes with a recommendation for mild reforms and an appeal to the better sense of all right-thinking zamindars. He displayed a firm conviction in the fact that there existed within the system both 'good' and 'bad' zamindars. He simplistically requested the former to motivate and pressurize their tyrannical counterparts to mend their ways. Two striking elements emerge from the author's handling of the peasant question. While he displayed a growing awareness about the grievous injustice meted out to millions of peasants there was also an underlying faith in the good intentions of colonial authorities. He was convinced that though consecutive legislation had strengthened zamindars, the ruination of the peasantry had never been the purpose. It was ignorance that had led to the errors in judgement on the part of colonial authorities. Ryots found it impossible to take recourse to law. The expenses and delay involved, as also the difficulties in proving their allegations hindered ryots.[103]

Bankimchandra was, however, not prepared to suggest any course of action that would disturb the existing superstructure of the Permanent Settlement as its abolition would cause a 'massive disorder' and a breach of promise on the part of the British government. Abolition of the system would dislocate the existing class structure based on the land system and cause total confusion in the social set-up of Bengal.[104] He voiced his dissatisfaction with Act X for having failed to restore amity between zamindars and ryots. It had failed to become a Magna Carta of the ryots and questioned the probity of the Act. It did not restore to ryots, he declared passionately all the rights they had lost since the Permanent Settlement. It also ensured that the ryots whose rent could not be hiked were insignificant in number. As a possible solution to the problem of exploitation of peasants, he advocated the extension of Permanent Settlement to ryots and a proper definition of the relationship between landlords and ryots.

His stand towards the peasant question saw distinct shifts over the years. Not in favour of a 'social revolution', he did not support the peasant agitation in 1873. He said that he was pained and disgusted by the conduct of the Pabna ryots. This is where he had serious differences of opinion with a section of the intelligentsia led by Surendranath Banerjee, the future founder of the Indian Association. The latter actively organized ryot unions in villages and lucidly articulated peasant grievances. In his review of the play *Jamidar Darpan* by Meer Masarraf Hussain in *Bangadarshan*, Bankimchandra expressed his disapproval at the conduct of the rebelling peasantry. While not denying the content of the play he advocated that its production and sale should be stopped at a time when the Bengal countryside was in turmoil.[105] This reflects the ambiguous attitude of the Bengali intelligentsia towards colonialism.

In another essay 'Samya'[106] published in 1879 the influence of Proudhon, St. Simon, Robert Owen, Fourier, Louis Blanc and John Stuart Mill is apparent. The author was inspired by contemporary attempts to remove inequalities among men. Here he analysed the economic basis of human civilizations and the hapless condition of the common men, the peasantry in particular as there was still no working class in the country. He advocated a total reform of the class-ridden social systems and establishment of a society based on equality.[107] Yet, he did not believe in removal of all inequalities and social divisions that could lead to a social revolution. He only advocated removal of excesses of all kinds. Seeking reasons for the socio-economic malaise afflicting the peasantry he blamed it on demographic changes and craving for wealth and its accumulation. But while he recognized the need to remove excessive inequality and eliminate class conflicts, he wanted no social upheaval. It was perhaps these contradictions in the essay that, unable to reconcile these convictions, he probably stopped any further reprints of the essay.[108]

In the meanwhile, the 1870s also witnessed for the first time the entry of Muslim writers in Bengal, hitherto more attracted to Urdu and Persian, into the domain of Bengali literature. This trend commenced with the publication of *Ratnabati* by Meer Masarraf Hussain in 1869.[109] The audience of Muslim-Bengali literature was far removed from more contemporary fictions and novels and were more inclined towards the supernatural and unreal, abounding in monsters, sorcery, adventures and intrigues. Nevertheless, Masarraf Hussain was an exception. He was the first to bring the peasant into the centre-stage of literary compositions. Contemporary journals and newspapers, viz., *Hafez* and *Kohinoor*, edited by Muslim litterateurs mostly dealt with themes related to tales of past glories of Muslims, Islamic religion and culture. On the contrary, a keen awareness of reality informed Masarraf's works.

His long association with the working of zamindari estates gave him an insight into the actualities of agrarian society. He worked for long as manager in an estate in Tangail. His short play *Jamidar Darpan* published in 1873 is a significant contribution to the cause of the long-suffering peasantry.[110] The background is the peasant revolt in 1872-3. It commences with descriptions of the oppressive nature of zamindars and the portrayal of general helplessness of the peasantry. The protagonist of the play, a poor peasant Abu Mollah, was forced to satisfy the whims of zamindar Haiwan Ali. Even his beautiful young wife Nurrunnaha is not spared. Victim of lust she is fatally injured in the process.[111] A remorseless Haiwan's only concern was how best to dispose off the corpse. The play portrays how law had become an instrument for zamindars to oppress their hapless tenants; the power structure was strengthened through links with the colonial judicial system with its overwhelming emphasis on evidence. Money often influenced proceedings and the rural poor found it impossible to defend their case. The oppressed and the exploited failed to put up any resistance. It is strange that a play written against the backdrop of the large-scale peasant unrest that swept the countryside failed to capture this expression of peasant consciousness, although Dinabandhu had brought peasant resistance into literature. The timing of the publication of the play is, however, significant.

It was at almost the same time that the *Peasantry of Bengal* by Romeshchandra Dutt was published. Along with Dadabhai Naoroji and M.G. Ranade, Dutt was to provide economic ideologies required and demanded by the emergent political consciousness. The latter was one of the earliest entrants into the Indian Civil Service.[112] His years of practical experience of India's agrarian system, of Indian poverty and famines gave him an expert's knowledge. Reasoning within the constraints of time, he criticized the colonial system even though he never pleaded for its replacement by an indigenous alternative. Written in the heyday of his official career, *The Peasantry of Bengal*, a text that is now almost forgotten, created a great sensation.[113]

Well aware that it would earn him the ire of the colonial authorities, he still had it published. His biographer tells us that an 'ardent and whole-hearted sympathy with the ryot is the key-note of the work'.[114] Dutt was emphatic that 'the true progress of this country will only commence when the poor Bengal ryot will be bettered in condition'.[115] Dutt portrayed the peasant during British rule as victims of zamindari oppression, antagonistic class relations, government apathy and absentee landlordism. The text commences with a review of the land system that pre-existed the British. The thoroughly exploitative zamindari system, he erroneously concluded, existed in Bengal from time immemorial. He located the cause of exploitation of peasants in the local climate that hindered hard work and the general character of the people which, made them thoroughly 'imbecile and incapable of resistance'.[116] Every official vested with authority, he observes, was likely to turn oppressive and tyrannical without evoking much active opposition. Yet, he says, among 'patriarchal nations', there was always to be found a feeling of 'reciprocal kindness which tempers the otherwise harsh relationship between masters and servants'.[117] Hence, though thoroughly oppressed under 'Hindu and Muhammedan rulers' peasants were better-off than at present. As to the impact of colonial intervention in agrarian society, he asserted: 'Four scores of years have rolled away since the Permanent Settlement,—what fruits have these eighty years of active legislation borne for the poor ryot? Commerce has thriven but commerce to them is practically forbidden and if agriculture has been extended, the zamindars and not the ryots reap the blessings.'[118]

He thus condemned the Permanent Settlement in 1874 as a 'prodigious blunder':

> Seldom in the annals of any country has hasty legislation been productive of effects so calamitous as the ill-conceived Permanent Settlement. On the head of Lord Cornwallis will rest the blame, that the extortion of the zamindars and their underlings has not to the present day ceased—that the ill-feeling between the ryot and his master has advanced with the advance

of years. On his Lordship's head will rest the guilt, that the most fertile source of revenue in a fertile country has been closed up for ever, that the extension of cultivation has increased, not the wealth of the cultivators but the number of a class of impoverished idlers, the zamindars with a two *anna* or one *anna* share of the ancestral estate. On his lordship's head rests the blame that we do not see the faintest glimmerings of rural civilization.[119]

The English, Dutt recorded, were unable to rid themselves of certain legal ideas that they imported from England or to adopt legislation suitable for the peculiar circumstances of the country. Their notion of the 'right of property', he said, proved fatal to all their good intentions and prevented any action for the relief of cultivators. He also voiced his dissatisfaction with Act X for having failed to restore amity between zamindars and peasants and all rights that the latter had lost. He was also critical of it for not extending security to non-occupancy tenants who were the most numerous.

Yet, the ryots in Dutt's portrayal were capable of protest when all hope seemed to be lost. He noted with satisfaction that the peasantry 'have at last felt their power, and have on occasions exerted it'.[120] It was a 'good sign' of the times. Recalling the indigo revolt, he wrote 'Fifteen years ago this sort of oppressiveness of the great Bengal Indigo Company of Nadia called forth a loud protest from the people in the shape of a general rebellion.'[121]

Commenting on the agrarian unrest in Pabna and neighbouring districts, he contended 'And yet it were idle to deny that the differences and disputes between zamindars and ryots are daily assuming a serious aspect. We are aware that the combining of the ryots *en masse* against their zamindars is at the present day rare occurrence.'[122]

Ryots, he wrote, were driven to despair by zamindars. They would never have combined against the zamindars, he was certain, had there been other alternatives. The risk that a ryot faced of being harassed and ultimately ruined if he fell out with his superior was so great that he could never have turned aggressor unless compelled by exploitation. On the contrary Dutt cited instances to show how at times when estates of benevolent zamindars were at risk of being sold for arrears, ryots spared no pains or sacrifices to prevent such eventualities. Under such circumstances, he affirmed, the rising of the people *en masse* in an entire district was certainly a singular phenomenon among a peasantry as mild as that of Bengal.[123]

The sole alternative left for the colonial government, assured Dutt, if it sought to ensure prevention of such future agitations, was a permanent settlement between zamindars and ryots and permanent restraints on powers of the former. In concrete terms, his reforms implied acceding to two main demands of ryots, viz., no further enhancement of rent and no ejection from land. Acceptance of the principle of fixed rents, he suggested, was the sole viable means for healing class antagonism. Dutt strongly pleaded the cause of second and third classes of ryots: 'Declare that all ryots have rights of occupancy and are entitled to every increase in the productive powers of lands that may henceforth accrue and a stimulus will be afforded for effecting improvements which even Asiatic indolence will scarcely resist.'[124]

Once the ryot was assured that his savings were his own, that permanent limits had been set to the claims of zamindars then, the motive of self-interest would make him prudent, provident and thoughtful. This would bestow the fruits of labour on the class that actually toiled for it. This, he felt, would create 'a more glorious' permanent settlement and form the Magna Carta of a 'worthy class' of the peasantry.

Dutt's ultimate admission that it would be best for peasants if zamindars could be 'turned adrift' and an end put to the 'oppression of centuries' is significant. But he accepted that it was not practicable. Nor was he in favour of the recognition of the principle of peasant proprietorship on practical grounds. He informs that 'Considered in the abstract, the system of peasant proprietorship may deserve the high encomiums bestowed on it by Sismondi, Mill and other political economists,—but it is a question entirely different as to whether such a system would suit the habits of people and conditions of life as existing in Bengal.'[125]

To adopt such a principle or to abolish of the Permanent Settlement would be he feared to convulse society.

Dutt was thus far from being a revolutionary in his ideas at this stage. He stood for reform and that too within the framework of the existing structure. He was for the introduction of permanent restraints on the powers of zamindars and his views extended no further. He erroneously believed that zamindars would one day, justify the faith Cornwallis had reposed in them. This hope was best expressed in his novel *Bangabijeta*, published in the same year. Here he drew a pen-portrait of an ideal zamindar. Indranath, the protagonist, was the son of a rich zamindar. Not interested in the pursuit of wealth, he devoted himself to the welfare of his tenants.[126] What Dutt probably knew, with his administrative experience, but refused to admit was that the Indranaths belonged to a bygone era.

The Peasantry of Bengal also raised for the first time issues that Dutt was to address more passionately in future, viz., famine and irrigation. For effective prevention of famines, the bane of the countryside, caused by frequent droughts Dutt advised extensive irrigation works at government initiative. This, he admitted, would assure much needed relief to the ryots and secure a part of the crops.[127] A fixed portion of the revenue need be set apart annually for such work. The government also needed to undertake maintenance of granaries at nominal expense. To stock these granaries, he estimated, with 2 million tons of rice in an ordinary year when the commonest rice sold at Rs. 1.8 or 1.4 per maund would cost the government a small sum. However, it would plentifully feed over half the population of Bengal for the entire period of four to five months between the sowing and reaping seasons. Thus, the severest of famines could be combatted.

Over the years, many of Dutt's views underwent significant changes. Nevertheless, when *The Peasantry of Bengal* was published it sparked off a prolonged debate among all sections of contemporary society. The pro-ryot sections welcomed it as it strengthened their stand on the agrarian question and their critique of Act X. Nevertheless, it was not accepted in official circles. His stand on agrarian unrest was regarded as most objectionable.[128]

Another classic account of peasant life, a story well told, published at about the same time was Rev. Lal Behari Dey's novel *Bengal Peasant Life* alternatively called *Govinda Samanta*. Dey was at the same time a teacher of English language and literature, an editor and an eminent writer. Born in a Vaishnab family, his early years were spent in abject poverty. However, he was fortunate to have been taught in a missionary school. He joined the church and started his working life as a missionary. He later joined the government education department and taught at different times at Berhampur Collegiate School, Berhampur College and Hooghly Mohsin College. His prolonged stay in the *mofussil* and rural areas gave him valuable experience of rural life.[129] A realistic piece of literature, the novel is a powerful account of peasant life, written in English and then published in Bengali. It was written in response to an award of £50 announced by Joy Krishna Mukherjee, the zamindar of Uttarpara, for the best novel to be written either in Bengali or in English illustrating the 'social and domestic life of the rural population and working classes of Bengal'.[130] The novel was sent for adjudication in 1872 but the announcement was delayed and the award was given to Dey in 1874.

The reader is informed that he 'is to expect here a plain and unvarnished tale of a plain peasant, living in this plain country of Bengal'.[131] The author gives a vivid description of the social and domestic life of Kanchanpur village in the Burdwan district of Bengal through a day-to-day account of a family of tenant farmers, the Samantas. Dey recreates the history of rural Bengal in the first half of nineteenth century, by narrating events that occur chronologically from the birth of Govinda Samanta, the protagonist of the novel, through his growing up years to his ultimate ruin. It is a story of alienation of land from the ryot, the actual tiller. The narrative opens with a vivid description of the village, its physical boundaries and the inhabitants.[132] The space occupied by different caste groups is described as also cultivated and waste lands. The peasant family round which the story centres belongs to the Ugra-Kshatriya or Aguri caste. The author follows the fortunes of the family, indebted to the zamindar and the moneylender, since the birth of Govinda.[133] There are lucid descriptions about life in the village, the peasant household, village deities, rituals practised, the village school, prominence of the zamindar in rural society and some of the service classes crucial to village life. However, the author relates to all these through Govinda's family. The text recreates the varied aspects of rural life. Even while remaining true to the main story there are descriptions of the manner in which problems related to village administration, law, and order were settled within the village, without reference to the higher administrative authorities and the control wielded by the zamindar on the petty administrative officials in the village. As the story unfolds, the reader is told about the functioning of the rural market; the sites of cultural and verbal exchanges; the agrarian techniques adopted by cultivators; types of crops grown and their distribution. By narrating the unspectacular happenings in Govinda's family since his birth the author gives insight into the kind of struggles that their lives waged. Like all others belonging to his station Govinda too inherited after his father's death the lands which he tilled. Nevertheless, like all others he too inherited the legacy of debt to which he himself added considerably in course of his life. It is the story of the tenant losing out his 'paternal acres' that constitutes the core of the novel.[134] It is agrarian tension that is all-pervasive and informs the thoughts and activities of all the characters. The novel reflects the various complexities associated with the process of colonial land legislations, the land tenancy legislations that initiated a process of absentee landlordism and displacements from land.[135] Dey's novel locates land as the site for rural tension. The text depicts the linkage between moneylending and landlordism as a major feature in the countryside and deals with the existing power structure.[136] As the story unfolds the reader is informed about the series of misfortunes that affects the peasant family. There is reference to the tyranny of petty officials, especially the local police official. The zamindar is depicted as an oppressor who used to his advantage the Brahmanical system, colonial administration and his own socio-economic status to exploit rural peasants and artisans.[137] The protagonist Govinda's plight was so miserable that even the rural moneylender, moved to pity, lent to heavily debted Govinda at a rate considerably lower than what he normally charged.[138]

Dey expressed relief that Act X, which he too mistakenly regarded as the Magna Carta of the peasantry, had initiated a great change in the status of Bengal ryots and effected his legal emancipation. Most importantly, it had abrogated the hated Regulations of 1799 and 1812. But unfortunately Govinda was ruined before its enactment. It took him a decade to pay off his debts to the moneylender. Soon misfortune struck again in the form of an epidemic in 1870 that affected Kanchanpur, Govinda's ancestral village. The peasant family was seriously afflicted. Though most recovered, Govinda's mother succumbed. All these prompted him to incur fresh debts from the moneylender.[139] Exhausted by repeated exertions, his end was drawing near. He succumbed to the terrible calamity in 1873, when the district of Burdwan was faced with a famine. His crops failed and he was forced to go to Burdwan town and hire himself out as a day-labourer in the relief works undertaken by the benevolent Maharaja Mahtap Chandra Bahadur of Burdwan. But the thought of his degradation continued to haunt him and broke his heart. His health visibly declined and one morning he was found dead in his miserable hovel, far from his home and those he loved.[140]

Dey's tale told in powerful yet simple language, offers tremendous insight into the rigours of peasant life in Bengal's villages. He portrays their ability to discern their exploitation by the village superiors as also, their utter helplessness to save themselves. One significant fact that emerges is that, the anger at the machinations of the local zamindar could not be channellized into the wider agrarian movement that swept over the countryside. It is a pointer to the limitations of peasant consciousness. However, while there is no indication of collective popular protest there was a perception of protest in individual terms, viz., Kalamanik's (Govinda's paternal uncle) thirst for revenge against the zamindar for ruining Govinda. On the other hand, there are also hints at upper class fear of possible popular protest against oppressors in future. The text also reveals an underlying admiration for the English. The focus is on different kinds of local exploitation.[141]

Thus by the 1870s, the peasant as a social category had made its place in a significant way in contemporary literature of Bengal-vernacular as well as Indian English. The intelligentsia in Bengal striving to emerge as leaders of society could not ignore them. The prolonged agitation of the peasantry, first against the exploitative system of indigo cultivation and then against zamindari exactions, prompted a growing awareness of the ills afflicting them. Agrarian issues were taken up in earnest by the native press—both pro-ryot and pro-landlord in their leanings. All these litterateurs were actively associated with the press. Knowledge of social reality informed their works unlike many of their contemporaries based in and around Calcutta. Their professions, which necessitated prolonged stay in the rural areas, fostered close association with the countryside.

However, their works were not essentially anti-colonial discourses and did not question British rule. They enjoyed the social pre-eminence their chosen professions afforded. Bankimchandra and R.C. Dutt were associated with government service and direct beneficiaries of social and economic opportunities offered by colonial rule. The same was the case with Dey, associated as he was with the government education service. Masarraf Hussain worked on estates of zamindars created by the British land system. The full implications of colonialism were not evident to them yet. They did not advocate a change in the land system or dismantling of the social structure based on the land system. They advocated reforms within the existing framework. They did not focus on the theme of peasant unrest. But the fact that these works were published when rural Bengal was engulfed by it indicates that contemporary happenings informed their psyche. Literary focus, increasingly progressive and realistic, had thus shifted from the historical and the romantic to real life themes. This trend, visible in fictional prose, novels, essays and all other forms of literary modes since the beginning of 1850s had reached a level of maturity by now. It had emerged as a mode of social protest as litterateurs were certainly motivated by a deep realization of the actual realities of peasant life. The protracted attempt of the government in this period to solve some of the sources of agrarian discontent no doubt gave an impetus to the publication of such works.

THE RENT QUESTION AND THE LEGISLATIVE PROCESS

On 9 June 1880, the Rent Law Commission submitted its report with a draft landlord and tenant bill. In the draft Rent Bill, the Commission desired to maintain the existing rule by which occupancy right was acquired by twelve years' continuous possession. The Commission also gave those ryots who had land for three or more, but, less than twelve years, a higher position and status than that of the tenants-at-will. The Commission, however, refused to give occupancy right to the *korfas* or under-tenants. The holding of occupancy ryots was declared transferable by private sale or gift and divisible by will but he was prohibited mortgage of his holdings.[142] The Commission also suggested that no ryot be ejected from land in which he possessed the right of occupancy whether for non-payment of rent or any other cause not being a breach of stipulation in respect of which, such ryot and his landlord had contracted in writing, but the ryot should be liable to ejection for a breach thereof. Henceforth if ryots were forced to relinquish their lands, zamindars were required to give them one year's enhanced rent as compensation for disturbance besides the value of their improvements.[143]

The Commission retained the three grounds of enhancement under existing law but it split the second ground into two parts: that the productive power of land has been increased otherwise than by the agency or at the expense of the ryots and from causes not merely temporary or casual and; that the price of produce has been increased otherwise than by the agency of the ryots and from causes not merely temporary and casual.[144] If improvement was made by the landlord only and not by the ryot, the landlord could enhance the rent and if neither of them was responsible for the increased value then the increment would be equally shared by landlord and tenant. The same procedure was to apply in case of price rise. The fair rent was also given a safeguard of a maximum limit beyond which rent could never be enhanced.[145]

The Government of Bengal, however, did not pass the Rent Bill of the Commission. Eden's government submitted in July 1881, another rent bill in which some of the pro-ryot clauses of the original bill were modified. This bill was further modified by the Government of India, which proposed classification of land instead of the status of the tenant as the basis on which the recognition of the occupancy right should be affected, and to attach the right to all ryoti lands.[146] But the Secretary of State opposed this proposal.[147]

On 2 March 1883, Sir Courtney Ilbert introduced in the Council a revised bill.[148] It was based on the original bill submitted by the Rent Law Commission.[149] Accordingly, it was provided that the right of occupancy could be secured by any ryot who held any ryoti land (whether the same or not it did not matter) in the same village or estate for the period of twelve years.[150] The occupancy ryot was given the right to transfer his holding, subject to a right of pre-emption in the landlord to have them at a price to be fixed by the civil court. The Bill also accorded some protection to the ordinary ryot from arbitrary rack-renting and eviction. Where the ordinary ryot refused to agree to an enhancement of rent the landlord could not eject him without paying him, besides any amount that might be due to him as compensation for disturbance, equal to certain multiple of the yearly increase of rent demanded.[151] Again the landlord could not eject an ordinary ryot except for arrears of rent; on the ground that the ryot had used the land in a manner which rendered it unfit; that he had refused to agree to an enhancement of rent. An occupancy ryot, whose holding consisted of ryoti land and who paid rent in kind, or paid as rent the value of a certain share of the actual gross produce of the holding, might, notwithstanding any contract on the contrary, sue to have his rent commuted into a money rent. The Bill proposed that money rent payable by an occupancy ryot might be enhanced by a contract in writing, approved of and registered by a revenue officer appointed by the local government on his behalf, and a revenue officer should not approve or register such a contract until he had satisfied himself that it was fair and equitable. It fixed the limits up to which the court was to enhance in cases, in which the enhancement was claimed on the ground of an increase in the productive powers. The enhancement was not in any case to be more than one-fifth of the established average annual value of the gross produce of the land in staple crops calculated, at the price at which the ryot sold at harvest time.

The rent of an occupancy ryot once enhanced, could not be increased for ten years. Where an occupancy ryot paid rent in kind or paid as rent the value of a certain share of the actual gross produce, the Bill provided that notwithstanding any contract or custom to the contrary, the landlord should not be entitled to recover, in respect of any portion of the holding on which staple crops were grown, more than half the gross produce or its value, as the case might be.[152] The ordinary ryot should pay such rent as might be from time to time agreed on between him and his landlord but subject to the provision that made 5/16 of the average annual value of gross produce the maximum of the rent.[153]

The Bill thus aimed to adopt a policy that would be fair to landlords and tenants alike even while seeking to remedy some of the defects of the existing law.[154] But it failed to create the right of occupancy in favour of the actual cultivators of the soil. Thus, the Bill was bound to create a set of occupancy tenants who were not the actual cultivators, but middlemen, and the latter would be in a position to oppress the cultivators.[155] Acquisition of occupancy rights was limited to lands in the same village only.[156]

On 14 March 1883, the Bill was sent to the Select Committee which submitted an amended bill on 29 March 1884. This Committee included pro-zamindar luminaries, viz., the Maharaja of Darbhanga, Kristo Das Pal and Ameer Ali. It recommended some modifications in the bill.[157]

The amended bill while not sufficiently protecting interests of tenants greatly facilitated enhancement of rents by specifying rise in prices as the ground for increase in place of requirement of proof of an increase in the value of the produce.[158] The landlords were to benefit further as the original safeguard that the enhanced rent should not in any case exceed one-fifth of the estimated average value of the gross produce of the land in staple crops, calculated at the price at which ryots sold at harvest time was not retained. Landlords' plea that the rate of rent paid by a ryot was below the prevailing rent was accepted as the justification for rent enhancement. The amended bill also did not limit the rent, which landlords could demand when a tenant was first admitted to the occupation of land. Even in the case of a settled ryot, the landlord could demand an increase of 25 per cent on the rent previously paid. The landlords' right to rent enhancement was strengthened to an unprecedented degree and they could extract the maximum amount that tenants would agree to pay rather than give up the land. The amended bill thus left the occupancy ryot without any adequate security in the matter of fair rent. It left ample scope for rack-renting without allowing any encouragement for cultivation of more valuable crops.[159]

The amended bill did not facilitate the acquisition of the right of occupancy by the non-occupancy ryot. It provided that he should not contract himself out of the power to acquire the right. But this safeguard was fruitless so long as the bill ensured that the right could never accrue. The right could be acquired only by twelve years continuous occupation of village land. As regards to rent, when first admitted to occupation the ryot must pay such rent as might be agreed upon between himself and his landlord. If he was afterwards called upon to agree to an enhancement he could demand a judicial rent for a term of five years. But this check was ineffective because unless he accepted the landlord's demand of rental rates, the landlord would not formally demand an enhancement, but would sue to have him ejected.[160] The landlord was thus empowered to demand a rent that would leave the tenant with the bare subsistence and the demand for enhancement might be repeated year after year. The bill thus put landlords in a position to rack-rent non-occupancy ryots during the eleven years, to evict them and begin the process again. The non-occupancy ryots were just tenants-at-will with only those rights that were allowed them by landlords. The mass of under-ryots were left totally unprotected. They could secure to themselves the same rights by taking registered leases. But the majority of them were too poor to demand registered leases.

The Bill that was finally approved by the Council and became law on 1 November 1885 was in fact a workable compromise.[161] The Act made acquisition of occupancy right an easier process. It provided that a ryot who had been in possession of any land for twelve years, either himself or through inheritance, would become a settled ryot of the village with occupancy right in the land he already possessed. This was an improvement on the existing law. However, fixity of tenure was limited by restricting the definition of the settled ryot to the village alone. The occupancy ryot could not be ejected except by court decree on the ground that 'he has used the land comprised in his holding in a manner which renders it unfit for the purposes of tenancy, or that he has broken a condition consistent with the provisions of this Act, and a breach of which is, under the terms of a contract between himself and his landlord liable to be ejected'. Yet, it afforded a remedy against absolute ejectment.[162] An occupancy ryot was not liable to ejection for arrears of rent.

The non-occupancy ryot continued to suffer. The gross produce limit of his rent was cancelled. He was declared liable to ejection on the grounds: that he had failed to pay an arrear of rent; that he had used the land in a manner which rendered it unfit for the purposes of the tenancy; that he had refused to agree to pay a fair and equitable rent or; that the term for which he was entitled to hold at such a rent had expired.[163] He could not claim a judicial rent for a period longer than five years. If the ryot agreed to pay the rent so determined, he would be entitled to remain in occupation of his holding at that rent for a term of five years from the date of the agreement but, on the expiry of that term would be liable to ejection subject to the provision of the Act unless he had acquired a right of occupancy.[164] Thus a non-occupancy ryot might be ejected at any time before he had acquired a right of occupancy without obtaining any compensation for disturbance.

As to the lender-ryots, it was admitted that they might have a right of occupancy by custom. It imposed a limitation on their rent of 50 per cent above the rent paid by their ryots to landlords where a registered case was executed. The period of leases to under-ryots was limited to nine years and they could be ejected on the expiry of the lease. The lender-ryots therefore remained unprotected.[165]

The Act VIII of 1885 thus put an end to the marathon rent controversy. It fell far short of the safeguards actually intended for the occupancy ryots by the Rent Law Commission. The provisions for under-ryots and non-occupancy ryots were not enough to secure them against the growing competition for land and left their fate to future settlement. Thus while the Act attempted to address some problems it failed to find a permanent solution to a set of contradictory interests.

REACTION OF LANDLORDS AND THE PRO-LANDLORD INTELLIGENTSIA TO CHANGES PROPOSED BY THE ADMINISTRATION, 1880-1885

K.D. Pal voiced the suspicious of zamindars and their supporters at the appointment of the commission as he retorted contemptuously:

Heretofore, the complaints were about the recovery of rent and the settlement of rent, but now came theory for a general revision of the rent law and a comprehensive revision of the substantive rights of landlords and tenants. . . .

That commission I believe was appointed with a view of securing the Bengal ryots that were popularly called the three F's (Fixity of tenure, Fair rent and Free Sale).[166]

'A Lover of Justice' expressed the fear that landlord-tenant relations, already strained since 1859, would further deteriorate.[167] When the recommendations of Rent Law Commission were made public in 1880 there was an outbreak of protest from zamindars who were scared of losing rights conferred by the Permanent Settlement.[168] Protest meetings were held all over Bengal to condemn the Bill.[169] In his speech before the Governor-General's Council K.D. Pal thundered:

I do not quite, understand the primary object of the Bill. Is it to prevent dispute and obligation and to promote peace and harmony amongst zamindars and ryots by reasonable fair and equitable provisions, or is it to redistribute rights in land, to promote and foster litigation, and set class against class. I should be very sorry to believe that the authors of the Bill have the latter object in view.[170]

Raja Narendra Krishna Dev, the president of the British Indian Association, Durgacharan Laha and Ashutosh Mukherjee, an eminent lawyer fighting the landlord cause shared the same view.[171] The legal authority of the Council to enact such a Bill was questioned.[172] The new bill was regarded as a greater threat to rights inherent in the absolute proprietor of land than the earlier Act X.[173] Maharaja Jyotindra Mohan Tagore, ex-member of the Vice Regal Council was even more articulate in his opposition. At a meeting of the British Indian Association held on 5 in April 1883 he suggested 'If however, the zamindars are considered an anachronism in the communistic spirit of the age . . . it would be better to purchase the zamindars out by paying them a fair market value for their estates.'[174]

On 17 November 1883 at a large public meeting held at the town hall in Calcutta, landlords unanimously condemned the proposed bill.[175] They feared that while there would be greater class tension in rural society,[176] the bill would not succeed in improving the condition of the actual cultivators who would always remain sub-tenants. They constituted the mass of the rural poor.[177] Some even feared that if the provisions relating to occupancy right proposed in the bill were enacted then a new class of petty sub-proprietors would emerge from among the existing tenants. They would in the long run transfer their lands to the baniyas, a fear that would be lucidly articulated decades later by Bengal's foremost poet Rabindranath Tagore.

The pro-zamindar press too took up cudgels against the revised bill of 1883. While the *Dainik Varta* viewed it as detrimental to the interests of both ryots and the country,[178] the *Reis and Rayyet* held further modifications in the Bill indispensable to serve the interests of all.[179] The *Charu Varta*[180] upheld these views while the *Rangpur Dik Darshan* was severely critical of the basic assumption of the government that all zamindars were equally oppressive.[181] It held the middlemen responsible for every act of tyranny practised on the tenants.

In the Select Committee, to which the bill of 1883 was sent, K.D. Pal dissented from the report of the majority members. Fearing that revision of the Rent Act would undermine the position of zamindars he objected to the bill.[182] Zamindars feared that the bill would benefit *jotedars* and middlemen and a class of small landowners would replace large zamindars.[183] They opposed any attempt at a Permanent Settlement between ryots and the government.

Zamindars tried to delay the Act even at the final stages. A sum of Rs. 35,000 was subscribed at a meeting held at the house of the Maharaja of Darbhanga on the eve of its enactment to organize a protest agitation in England.[184] The clauses were condemned as 'revolutionary' intended to create unrest and infringing on the guaranteed rights of zamindars.[185] Fear of ruination of both zamindars and ryots, continual litigation and rural disturbances was voiced in the press.[186] The pro-landlord press carried on a spirited protest to prevent enactment of the Bengal Tenancy Act. When it was passed in spite of all opposition, the pro-landlord intelligentsia was thoroughly dissatisfied with the outcome.

PRO-RYOT INTELLIGENTSIA AND THE LEGISLATIVE PROCESS

In the decade preceding enactment of the Bengal Tenancy Act in 1885, the question of peasant right had become a hotly debated issue among the urban intelligentsia. Inspired by Benthamite ideas and protagonists of peasant interest, a progressive section began to question the domination of zamindars in rural society. They were primarily service-holders or belonged to the different professions.[187] The Indian Association took the lead in championing the cause of peasants during the controversy over the Rent Bill. Members campaigned for pro-ryot changes through meetings attended by hundreds of peasants and Rent Unions. They decried the demands and proposed laws in favour of occupancy ryots. Most members of the Association were products of Hindoo College. To broaden the base of the former the proposal of its assistant secretary Dwarkanath Ganguli to reduce the general membership fee from Rs. 10 per annum to Re. 1 for the ryot members was accepted in 1880. The Association soon emerged as an adversary of the British Indian Association, which was a representative of the interests of the landed class. However, the conflict between the two was never fundamental. It was perceived as one between an old landed class trying to hold on to their rights and a gradually growing middle class that was not based, unlike their European counterparts, on business and commerce. The latter too had landed roots. The enactment of the Bengal Tenancy Act and the inception of the Indian National Congress cemented the clash between pro-ryot and pro-zamindar sections of the intelligentsia. The members of the British Indian Association and the Indian Association came together to serve the greater political cause.

On the issue of the recommendations of the Rent Law Commission, Syed Ameer Ali was the sole Indian member in the Legislative Council to raise his voice on behalf of the ryots.[188] He said in the Council that it seemed that the government had 'abdicated in perpetuity its legislative function to protect the numerous class of peasantry'.[189] He was in effect expressing the opinion held by the advocates of peasant rights. They welcomed the recommendations of the Rent Law Commission. The Indian Association organized for the first time public meetings of ryots at Rohouta, Kissengunge, Goospara and Gopalpur in the Nadia district; at Logusai in Birbhum; at Rahita in the 24-Parganas; at Baidyabati in Hooghly; at Burdwan and at Calcutta. Ryots were thus motivated to struggle for their own cause.[190] At Kissengunge in a meeting attended by hundreds of ryots, an appeal was made for the enactment of the Bill.[191] In favour of the Rent Bill the Association welcomed the government's decision to legally settle relations between landlords and tenants.[192] It suggested that occupancy tenure be made transferable and mortgageable.[193] It also proposed that the prescriptive right for acquiring occupancy status be reduced for three years and a maximum limit for enhancement be imposed.[194]

However, in their handling of the clauses of the Bill the liberal and pro-ryot section of the native press displayed rather contradictory attitudes. While the *Bengalee*[195] was convinced that the Rent Bill once enacted would ensure great benefits for the peasantry the *Som Prakash* feared the worst for both zamindars and ryots. Nevertheless, the attempts of the zamindars to block the proposed Bill in 1883 invited a storm of criticism. The *Burdwan Sanjivani* condemned all such attempts while the *Bengalee* deplored zamindars for negotiating with Tory leaders in England to stop its enactment on the ground that 'behind the interest of the class or the party, are the thrice-seared interest of our country'.[196] The pro-ryot press was also critical of the shortcomings of the Tenancy Bill. The *Bengalee* regarded the exclusion of the *khas mahals* from the operation of its provisions and the failure to provide adequate safeguards for ryots in indigo districts as the gravest defects.[197]

At the next stage in the legislative process, when the Select Committee submitted the amended Bill in 1884 they were quick to articulate their views. The proposed changes left them dissatisfied. They held that the amended Bill brooded ill for the peasantry, as it did not provide them the security of tenure that had been the original objective. The zamindars on the contrary, were assured easy facilities for enhancement. The *Sanjivani* too was critical of that section which allowed them to enhance rent if they could prove that ryots' income from sale of the produce of the land had increased even when that increase was not due to any investment or efforts undertaken by zamindars.[198] The *Som Prakash* in the 1880s dissented from the *Bengalee* and objected to the conferring of the right of transferability on occupancy ryots on the ground that ultimately zamindars would buy them out and would rent out these lands to ordinary ryots. The occupancy ryots would soon disappear from Bengal.[199]

In the Council, Ameer Ali carried on valiantly his crusade against zamindars and their supporters. However, unlike *Som Prakash* which was against extension of the right of occupancy[200] he advocated that it be conferred on non-occupancy ryots and under-ryots.[201] He demanded free transfer of occupancy holdings and endowing occupancy ryots with the right in Bengal proper to transfer holdings in the manner similar to other immovable property. The landlord, however, was to be entitled to a fee of 10 per cent of the purchase money.[202] He opposed all attempts by zamindars to amend provisions relating to the right of transfer enjoyed by occupancy ryots.[203] He objected to the provision of enhancement of rent on the ground of increase in the price of staple food crops.[204] Ameer Ali was supported by a powerful section of the press. The *Shakti* and the *Sambad Bahika* looked forward to a speedy solution to the rent question to avoid any 'great commotion' in future.[205]

The *Bengalee* did not share the apprehensions of the zamindars.[206] On the contrary, the Indian Association appealed in its letter to the government of Bengal for a precise definition of tenure and that of a ryot to prevent further class friction in agrarian society. It particularly sought provisions to prevent ensuing confusion between different classes of tenants.[207] As regards the issue of enhancement of rent the Indian Association preferred the tenant to be left free to contract up to the maximum limit to which rent could be enhanced in the court. However, it advocated that tenants be debarred from accepting any increase within ten years from the date of the last enhancement. While the revenue authorities could be entrusted with the preparation of a price list for food grains, the fixation of equitable rents was to be left to the civil courts. The Association was also in favour of depriving zamindars of the power of distraint as the provisions in the Civil Procedure Code for attachments before judgement in case of rent suits could suitably protect zamindars against any machinations of tenants.[208]

The Indian Association protested against revisions made in the Tenancy Bill by the Select Committee which if implemented could disastrously affect the interests of ryots and aggravate agrarian tension. It objected to the Committee's recommendation that occupancy right be no longer transferable and that transferability be regulated by custom.[209] It advocated that in Bengal occupancy right be transferable by sale. It criticized the Select Committee for omitting provisions that entitled non-occupancy ryots to compensation upon unjust eviction. This would frustrate the avowed object of the Bill to enable non-occupancy ryots to acquire gradually the position and status of occupancy tenant. The Association also feared that the revised Bill if enacted would be detrimental to the condition of under-ryots.[210] It proposed compensation for them in case of their dispossessions.

THE STORM BLOWS OVER

None of the objections of the pro-ryot literati were considered by the government. At the intervention of Lord Dufferin, the Council made considerable concessions to zamindars in contrast to the provisions of the original Bill.

The Act when passed left much to be desired. However, a section of the pro-ryot literati was temporarily silenced. R.C. Dutt was convinced that the ryots were saved from undue enhancements, the average rent paid by them did not exceed one-sixth of the gross produce in any district and was much lower in the eastern districts.[211] The Act, he said, had effectively succeeded in protecting ryots against all unjust enhancements and evictions.[212] It was thus after the labours of a century that the British administration, he felt, had solved satisfactorily the 'land question' in Bengal firstly by extending protection to zamindars by the Regulation of 1793 and secondly by extending protection to actual cultivators by the Acts of 1859 and 1885. Reviewing the consequences more than a decade after its enactment he observed that cultivators in Bengal were more prosperous, more self-reliant and safer against the worst effects of famine, than cultivators' elsewhere.[213]

NOTES

1. Oppression of our weak ryots
   Zamindars oppress so many
   They know in their hearts that without zamindars
   There is none to listen to their woes
   Peasants suffer without saying anything
   Think in their hearts there is no way out
   Zamindars often fine them
   Oppressing peasants to their heart's content
   Listen gentlemen pay attention
   Show you its dramatization today.
   Meer Masarraf Hussain, 'Jamidar Darpan', in Bishnu Bose (ed.), *Meer Masarraf Hussain Rachanasamgraha*, vol. I, Mitra & Ghosh, Calcutta, 1978, pp. 42-3 (translation mine).
2. Sanjibchandra Chatterjee, *Bengal Ryots: Their Rights and Liabilities*, Pub. n.m., Calcutta, 1864.
3. B.H. Baden Powell, *The Land Systems of British India*, vol. I, Oxford University Press and Steven & Sons Ltd., London, 1892, pp. 644-5.

4. Radharaman Mukherjee, *Occupancy Rights: Its History and Incidents*, University of Calcutta, Calcutta, 1919, p. 79.
5. *Report on the Land Revenue Commission*, 1940, vol. I, p. 25.
6. B.B. Chaudhuri, *Growth of Commercial Agriculture*, op. cit., p. 202
7. By the Regulation 5 of 1830, a peasant found guilty of breach of contract with the indigo planter, viz., failure to cultivate indigo, would be sentenced for a period not exceeding one month. Moreover, the Magistrate could 'require' the convicted peasant to sow or cultivate the land; a disobeying peasant would come in for a more rigorous punishment of a longer period of imprisonment not exceeding two months. (Ibid., p. 158.)
8. *Report of the Indigo Commission*, Calcutta, 1860. Evidences of Larmour, Herschel, Forlong, A.P. Mclean, A. Hills.
9. Asoka Kumar Sen, *The Popular Uprising and the Intelligentsia: Bengal Between 1855-1873*, Firma KLM, Calcutta, 1992, pp. 60-1.
10. *Hindoo Patriot*, 4 February 1860.
11. Rev. C. Bamwetsch wrote letters to the *Hindoo Patriot* denouncing the conduct of the men of his own race. Ref. Asoka Kumar Sen, op. cit., p. 61.
12. I agree with Asoka Kumar Sen that the facts about indigo oppression from the actual scene of occurrence rendered the press coverage authenticity as well as greater credibility. Ibid.
13. Dinabandhu Mitra, 'Nil Darpan', *Dinabandhu Rachanavali*, Sahitya Samsad, Calcutta, 1967.
14. *Hindoo Patriot*, 19 May 1860.
15. Ballads were sung in the countryside in his praise:
    'Our Harish saved us while another Harish (a servant of an indigo factory) killed (us).' *Report of the Indigo Commission*, Evidence, op. cit., pp. 45-7; Asoka Kumar Sen, op. cit., p. 70.
16. *Hindoo Patriot*, 1 August 1860.
17. Selections from the Records of the Government of Bengal, no. XXXIII, part I, pp. 10-25; Asoka Kumar Sen, op.cit., p. 71.
18. *Bengalee*, 28 October 1873; *Friend of India*, 16 October 1873.
19. Letter from the Secretary, British Indian Association to F. Clarke, Offg. Secretary to Government of Bengal dated 12 June 1875 (British Indian Association Records), Progs. of the Annual Meeting, dated 7 and 29 April 1874, Circular dated 8 June 1875 (British Indian Association Records), cited in K.K. Sengupta, *Pabna Disturbances and the Politics of Rent 1873-1885*, People's Publishing House, New Delhi, 1974, p. 118.
20. *Permanent Settlement Imperilled or Act X of 1859 in its True Colours* by a Lover of Justice, Calcutta, MDCCC LXV, 1865.
21. *Bengal Land Revenue Proceedings*, nos. 14-52, 14 February 1877.

22. Ibid., no. 2, 1861.
23. Progs. of the British India Association's annual meetings, dated 7 and 29 April 1874.
24. Radharaman Mukherjee, op. cit.; 'Home Department (Police) Proceedings', p. 87; August 1873, pp. 24-9, vide M. Chattopadhyay, Petitions, p. 35.
25. *Sulabh Samachar*, 2 September 1874.
26. *Bengalee*, 5 July 1873.
27. *Hindoo Patriot*, 13 October 1873.
28. *Bengalee*, 5 July 1873.
29. Although the leading historian of the Pabna disturbances K.K. Sengupta denies the official view that the prosperity of the ryots was a causal factor in the tensions, according to Sugato Bose's analyses the battle was in part about the share-out of the higher gross income brought in by jute. Sengupta, op. cit.; 'Agrarian Disturbances in the Nineteenth Century Bengal', *Indian Economic and Social History Review*, vol. 8, 1972, pp. 192-212; Sugato Bose, *The New Cambridge History of India, Peasant Labour and Colonial Capital: Rural Bengal Since 1770*, Cambridge University Press, Cambridge, 1993, p. 157.
30. K.K. Sengupta called it a protest against 'high landlordism'. He has defined it as a stage of change in the pattern of ownership and its impact on the tenantry that developed under nineteenth-century colonization, and of the growth of a desire on the part of some landlords to hold their estates untrammelled by even conventional rules and regulations for protection of their tenantry. Sengupta is of the opinion that these agrarian protests demonstrated the power of agricultural unionism that developed amongst certain sections of the cultivators, underlined the basically unstable nature of landlord-tenant relationship and pointed to the inadequacy of the existing law which governed these relationships. In the anti-rent agitation even though most landlords in Pabna were Hindus who formed just below 10 per cent of the district's population and of the peasants 70 per cent were Muslims, the religious factor never surfaced. However, it is a fact that peasants found it easy to combine together as the majority were Muslims. The Muslim society was not subject to caste dissensions that were so common among their Hindu counterparts. 'Report on the origin, nature and probable consequences of agrarian combinations in east Bengal' from J.G. Charles, dated 19 September 1873, misc. colln. 14, nos. 26-7, *Bengal Land Revenue Progs.*, January 1874; K.K. Sengupta, *Pabna Disturbances*, op. cit., p. vii.
Late nineteenth century saw a redefinition of caste and religious identities. Between 1872 and 1911 several castes changed their nomenclature and claimed higher rituals status: the kaibartas of west Bengal became mahishyas; the chandals of east Bengal namasudras, and; the koches of

north Bengal rajbansi kshatriyas. Bound to moneylending landlords and traders by their credit needs the Muslim peasants of east Bengal too showed signs of a developing awareness towards the closing stages of the century. However, I agree with B.B. Chaudhuri that on the whole the peasant movements except in some cases, viz., the Chandal movement in Faridpur in the 1870s were rarely organized on caste lines during this period and the new caste solidarity movements only marginally affected them. Chaudhuri, 'The Transformation of Rural Protest in Eastern India, 1757-1930', Presidential Address, Modern Indian History section, 40th session, Indian History Congress, Waltair, December 1979, p. 37.

31. The *Amrita Bazar Patrika* remarked:
'In the country in which from the immemorial up to twenty years ago, the landlord and the tenant were as father and son . . . the Penal Code and Act X have destroyed the cordial relation between the landlord and the tenant.' The *Amrita Bazar Patrika*, in *Bengalee*, 5 July 1873.
Again the *Truth* observed:
'The fact is that the law governing the ryots and the landholder is extremely defective and quite different from the spirit of the age. It is absurd to expect with this state of law to see the relation of zamindar with his lands and tenants placed on a satisfactory footing.' *Truth* in *Bengalee*, 6 September 1873.
Refer to Asok Kumar Chatterjee, 'Influence of Tenancy Legislation in Bengal's Economy 1858-1939', unpublished Ph.D. thesis, Rabindra Bharati University, 1986. This is a highly informative work on the evolution of land acts in Bengal. It is one of the earliest works on the subject. A wide range of contemporary journals have been sourced to bring out the different issues raised by contemporaries. I have found it extremely helpful for the analysis of the public debates on clauses of the tenancy bills.
32. *Sahachar,* 16 February 1874 (Report on Native Papers).
33. *Hindu Ranjika*, 31 December 1870 (Report on Native Papers).
This journal started as a monthly.
34. File 448, no. 37, Bengal Judicial Proceedings (Police), July 1873.
35. *The Gazette of India*, 1883, Extraordinary, p. 130.
36. Ibid.
37. C.E. Buckland, *Bengal under the Lieutenant Governors,* vol. II, Calcutta, 1901, pp. 640-1.
38. *The Gazette of India*, 1883, Extraordinary, p. 130.
39. Report on the Rent Law Commission, vol. II, p. 210.
40. Ibid., Land Revenue no. 229T, 15 July 1880, vol. I, p. 1.
41. Ibid., vol. I, p. 210.
42. Speeches of the National members of the Governor-General's Legislative Council on the Bengal Tenancy Bill, Appendix, pp. i-ii.

43. *Som Prakash*, 1 Ashar 1271 B.S., no. 31 (Report on Native Papers); *Sadharani*, 11 July 1875.
44. *Bengalee*, 25 October 1873.
45. Ibid., 9 August 1873.
46. *Reis and Rayyet*, 11 August 1883 (Report on Native Papers).
47. *Sulabh Samachar* observed:
    'The object of the Regulations framed by the British Government for the good of ryots is being frustrated by the machinations of selfish zamindars.' *Englishman*, 8 July 1873.
48. Ibid., 2 September 1873.
49. *Sadharani* was published from Chinsurah. Bankimchandra Chattopadhyay and Jogendrachandra Basu were associated with this.
50. He was author of *The Mutinies and the People: The Career of an Indian Princess*, etc.
51. *Sadharani*, 9 May 1875.
52. *Bengalee*, 1 March 1879.
53. Biman Bihari Majumdar, *History of Indian Social and Political Ideas*, Bookland Pvt. Ltd., Calcutta, 1967, p. 98.
54. *Som Prakash*, 4 Aswin 1271 B.S., no. 45 (Report on Native Papers).
55. *Report of the Rent Law Commission*, 1880, vol. I, p. 30.
56. 'Letter of Humanities', *Hindoo Patriot*, 9 May 1857.
57. *Sadharani*, December 1873.
58. *Pioneer*, 16 July 1873.
59. Not only were new occupancy claimants required to furnish the proof that he had held every particular field of his holding for 12 years, zamindars could easily evade the law by shifting his tenant from one field to another to prevent its accrual.
60. *Bengal Land Revenue Progs*, no. 14-70/71, May 1879.
61. *Desh Hitoishini*, 1 September 1873 (Report on Native Papers, 13 September 1873).
62. *Som Prakash*, 14 July 1873.
63. Ibid., 18 Agrahayan 1279 B.S., no. 3 (Report on Native Papers).
64. Popularly known as Kangal Harinath, he was born in a respectable *tili* family in Kumarkhali village in Nadia. He was orphaned in childhood and could not complete his education because of poverty. All his life he struggled for the cause of the poor and exploited villagers. His entry into journalism was made possible by Ishwarchandra Gupta and *Sambad Prabhakar*. He also wrote for *Hindoo Patriot* for a brief spell. In 1863 he started *Grambarta Prakashika* and was associated with it till its publication was stopped because of shortage of funds. In adult life he came under the influence of Brahmo leader Bijoy Krishna Goswami. After 1884 he became involved with spiritual pursuits. He composed many *baul* songs. He authored 18 books, viz., *Bijaybasanta*, *Kabita Kaumudi*, etc.

65. 'Atyacharita Praja', *Grambarta Prakashika*, Falgun 1279 B.S./February, 1st week, 1873.
66. 'Jamidar O Praja (Bangla Pradesh)', ibid., Asar 1279 B.S./June, 1st week, 1872.
67. 'Jamidar', ibid., Asar 1279 B.S./June, 2nd week, 1872.
68. 'Praja, Jamidar O Government', ibid., Asar 1279 B.S./ July, 4th week, 1872.
69. 'Editorial', ibid., 1279 B.S./ September, 4th week, 1872.
70. Brojendranath Bandyopadhyay, 'Harinath Majumdar', *Sahitya Sadhak Chritmala*, no. 35, Bangiya Sahitya Parishad, Calcutta, 1350 B.S.; Sunil Bandyopadhyay, *Kangal Harinath: Sangbadikatay Anirban Angikar*, Samatat Prakashani, Calcutta, 1999; Prabir Kumar Debnath, *Prasanga: Kangal Harinath*, Mondol & Son, Calcutta, 1989; Amar Dutta (ed.), *Kangal Harinather Grambarta Prakashika*, Mitra & Ghosh, Calcutta, 1397 B.S.; Jaladhar Sen, 'Harinath Majumdar', *Dashi*, June 1896, p. 310. Cf. Jaladhar Sen was a contemporary, friend and biographer of Harinath.

    Harinath wrote his diary in 8 volumes numbering more than 2,000 pages. His biographers, viz., Jaladhar Sen, Brojendranath Bandyopadhyay and others have all read them. However, for reasons not known, the diary has not been published and at present cannot be traced. However, some portions have been published titled 'Kangal Harinath Aprakashita Diary', *Chatushkon*, Saradiya, ed. 1370 B.S. and Asar 1371 B.S.
71. 'Ryot Hungama', ibid., 1 Jaisthya 1283 B.S./ 13 May 1876.
72. 'Editorial', ibid., 8 Bhadra 1280 B.S./ 23 August 1873.
73. 'Kharap O Bhalo Zamindar', ibid., 30 December 1873.
74. Sengupta, op. cit., p. 122.
75. *Som Prakash*, 13 October 1873 (Report on Native Papers).
76. The *Sulabh Samachar* was echoing Harinath when it enjoined the government to compel zamindars to fulfil their obligations to ryots under the Permanent Settlement. *Sulabh Samachar*, 1 September 1873 (Report on Native Papers).
77. *Rajshahi Samachar*, 23 June 1875 (Reports on Native Papers).
78. *Bengalee*, 5 July 1873.
79. Buckland, op. cit., vol. II, p. 637.
80. *Som Prakash*, 5 March 1875.

    Cf. The Act VIII of 1869 clearly laid down the conditions under which the rent of an occupancy ryot would be enhanced. It transferred rent cases from the revenue courts to the civil courts, which made it difficult for zamindars to realize rent. Rigid observance of legal formalities in the civil courts deprived zamindars from acquiring any benefit out of rent suits. The 1876 Act transferred rent cases from civil courts to the collectors.
81. J.C. Bagal, *History of the Indian Association*, op. cit., Appendix, p. X.

82. *Dacca Prakash*, 26 December 1875 (Report on Native Papers). It was a weekly published from Dacca since March 1861. It was edited by the poet Krishnachandra Majumdar and later by Prasannakumar Bhoumik.
83. *Bengalee*, 2 May 1879.
84. *Som Prakash*, 5 March 1875 (Report on Native Papers).
85. *Amrita Bazar Patrika*, 20 May 1875.
86. *Sahachar*, 21 March, 10 May 1875; *Barisal Bartabaha*, 12 May 1875; *Suhrid*, 8 June 1875; *Hindu Ranjika*, 6 May 1875 (Report on Native) and many others. Cf. *Barisal Bartabaha* was the first Bengali newspaper to be published from Barisal. It was a fortnightly edited by Ishwarchandra Kar who resided in Magura village.
87. *Bengalee*, 8 March 1879.
88. Ryots complained that 'looking at the mass of papers attached to the bill, there does not appear any urgent necessity for such stringent and exceptional provisions in the law as are sought to be introduced in the bill'. *Bengal Land Revenue Progs*, no. 14-70/71, May 1879.
89. *Bengalee* wrote:
    'We congratulated Sir Ashley Eden in not having persevered with the Bill and on his having appointed a commission which will go into the question carefully and minutely and end their labour by presenting a bill on the subject.' *Bengalee*, 26 August 1879.
90. Herman Braet, 'A Thing Most Brutish: The Image of the Rustic in Old French Literature', in Del Sweeney (ed.), *Agriculture in the Middle Ages: Technology, Practice and Representation,* University of Pennsylvania, Philadephia, 1995, p. 191; Del Sweeney, 'Introduction in ibid., p. 3.
91. Sanjibchandra Chattopadhyay, *Bengal Ryots: Their Rights and Liabilities: Being an Elementary Treatise on the Law of Landlord and Tenant*, Calcutta, 1864, ed. A.C. Banerjee and B.K. Ghosh, rpt., K.P. Bagchi & Co., Calcutta, 1977.
92. His works include 'Palamau', 'Kathamala', 'Madhabilata' and 'Rameshwarer Adrishta'.
93. Sanjibchandra Chatterjee, op. cit., Chapter II, section I.
94. Ibid., p. 42.
95. Ibid., Chapters III, IV, VI.
96. Ibid., Chapter VII.
97. Bankimchandra himself edited the *Bangadarshan*. It was immensely popular among the literati.
98. Bankimchandra Chattopadhyay, 'Bangadesher Krishak', *Bankim Rachanavali*, vol. 2, Sahitaya Samsad, Calcutta, 1361 B.S., p. 309.
99. Ibid., p. 310.
100. Partha Chatterjee, *Nationalist Thought and The Colonial World: A Derivative Discourse*, Zed Books, London, 1986, p. 63.
101. S.N. Mukherjee and M. Maddern (tr. & ed.), *Bankimchandra*

Chatterjee: Sociological Essays, Utilitarians and Positivism in Bengal, Riddhi, Calcutta, 1986, p. 180.

102. Ibid.
103. Bankimchandra also mentioned that law courts being situated at considerable distance from interior villages and fear of retribution were great deterrents. Besides the ignorance of English judges about actual conditions and the inadequate number of courts delayed justice.
104. Bankimchandra Chattopadhyay, op. cit., p. 298.
105. *Bangadarshan*, 2nd year, no. 5, Bhadra 1280 B.S.
106. Published in 1879.
107. Bankimchandra Chattopadhyay, 'Samya', *Bankim Rachanavali*, op. cit., vol. 2, pp. 388-90.
108. Excessive socio-economic inequalities he felt could be reduced through good intentions of individuals and the intelligentsia were perhaps best equipped for the task. Yet Bankimchandra was rather scornful of this class of men.
    Bankimchandra Chattopadhyay, 'Lok Rahashya' and 'Kamalakanter Daptar', ibid., pp. 1-112. Partha Chatterjee, op. cit.
109. Masarraf began his career as a journalist and essayist in *Prabhakar* and *Grambarta Prakashika*. Between 1869 and 1911, he published 36 works. They include prose, poetry, songs, novels, plays, essays, stories, autobiographical works and others like school textbooks. A list has been compiled in Kazi Abdul Mannan (ed.), *Masarraf Rachana Sambhar*, vol. I. Bangla Academy, Dacca, 1976, along with tentative dates of their publication. Most are lost. Only about twelve full works have been located. Masarraf edited the fortnightly journal *Ajijan Nehar* from 1874 onwards. This is probably the oldest Muslim journal. Most of the issues are untraceable today.
110. Meer Massaraf Hussain, 'Jamidar Darpan', op. cit.
111. Ibid.
112. He served as a district officer in several districts and finally, for short spells as Commissioner of two divisions.
113. It was published in 1874.
114. J.N. Gupta, *Life and Work of Romeshchandra Dutt*, E.P. Dutton & Co., New York, 1911, pp. 55-60.
115. R.C. Dutt, *The Peasantry of Bengal*, rpt., ed. Narahari Kaviraj, Manisha Granthalaya, Calcutta, 1980, p. 95.
116. Ibid., p. 20.
117. Ibid.
118. Ibid., pp. 52-3.
119. Ibid., pp. 42-3. He ascribed the deteriorating zamindar-ryot relationship to colonial rule that made the landed sections significantly powerful. No efforts were made to check zamindari exaction or to define the rights of cultivators.

120. Ibid., p. 138.
121. Ibid., p. 63.
122. Ibid., p. 110.
123. Ibid., pp. 65-8.
124. Ibid., p. 73.
125. Ibid., Preface, pp. 3-4.
126. R.C. Dutt, *Bangabijeta*, Calcutta, 1874.
127. R.C. Dutt, *The Peasantry of Bengal*, op. cit., pp. 47-8.
128. J.N. Gupta, op. cit., pp. 55-6.
129. Dey's *Folk Tales of Bengal* is the first such work produced in the Bengali language. He was editor of the highly popular *Bengal Magazine* and *Friday Review* for considerable time. *Bharatkosh*, vol. 5, Bangiya Sahitya Parishad, Calcutta, 1973, p. 452.
130. Lal Behari Dey, *Bengal Peasant Life*, Macmillan & Co., London, 1934 edn., Preface, p. vii.
131. Ibid., p. 4.
132. Ibid., Chapter III, pp. 10-16.
133. Ibid., p. 49.
134. Ibid., Chapter XXIV, pp. 162-4.
135. Ibid., Chapter III, pp. 10-16.
136. Ibid., pp. 280-1.
137. Ibid., p. 361.
138. Ibid., p. 367.
139. Ibid., pp. 369-74.
140. Ibid., pp. 375-6.
141. This realistic piece had a profound impact on contemporary literature. It exerted for instance a shaping influence on the first Oriya novel *Cha Mana Atha Guntha* written by Fakir Mohan Senapati, a quarter of a century later. Himansu Sekhar Mahapatra and Jatindra Kumar Nayek, 'Writing Peasant Life in Colonial India: A Comparative Analysis of Lal Behari Dey's *Bengal Peasant Life* and Fakir Mohan Senapati's *'Cha Mana Atha Guntha'*, *Jadavpur Journal of Comparative Literature*, vol. 32, 1994-5, p. 22.
142. *Report of the Rent Law Commission*, vol. I, p. 15, para 28.
143. Ibid.
144. Ibid., pp. 29-31.
145. Section 23 laid down that when the rent of an occupancy ryot was enhanced on the basis of the first, third or fourth ground, the enhanced rate should not be more than one-fourth of the average value of the gross produce. The Commission said that when the rate payable by a ryot for an occupancy holding was paid in kind and the ryot received no assistance from his landlord towards producing the crop such rent should not exceed 50 per cent of the gross produce in the case of staple crops; that in the case of special crops, if the landlord and tenant

had earlier contracted in writing for any particular share of the gross produce, in such contract the landlord should not be entitled to recover any other or greater rent than 25 per cent of the average annual value of the gross produce.
Ibid.

146. Ibid., p. 84.
147. The Government of India decided that it would not introduce a bill in the form which was disapproved of by the Secretary of State. It was determined that the measure should be framed upon the lines suggested in the latter's despatch.
Ibid., p. 809.
148. Ibid.
149. *The Supplementary Gazette of India*, 21 April 1883.
150. *The Gazette of India, Extraordinary*, 1883, p. 134.
151. Ibid., p. 141.
152. Ibid., p. 102.
153. Section 119, *The Gazette of India*, 29 March 1884, p. 30.
154. *The Supplementary Gazette of India*, 21 April 1883, p. 867.
155. Ibid., pp. 832-3.
156. Radharaman Mukherjee, op. cit., p. 99.
157. *The Gazette of India*, Part V, 29 March 1884, p. 26, for details of the Bill.
158. Ibid., 12 April 1884, p. 276.
159. *The Extra Supplementary Gazette of India*, 14 February 1885, p. 3.
160. Ibid., p. 22.
161. Radharaman, Mukherjee, op. cit., p. 98.
162. Baden-Powell, *The Land System of British India*, vol. I, op. cit., p. 652.
163. Ibid., section 47.
164. Ibid., section 46.
165. *Report of the Land Revenue Commission*, vol. I, p. 30.
166. *Speeches of the Native Members of the Governor General's Legislative Council*, p. 17.
167. *'A Lover of Justice'*, op. cit., p. 29.
168. *Bengalee*, 13 November 1880.
169. Meetings were held in Calcutta, Dacca, Bankipur and elsewhere. *Som Prakash*, 27 Pous 1287 B.S., no. 9 (RNP); *Bengalee*, 18 December 1880.
170. Speeches of the Native Members, op. cit., p. 17.
171. Raja Narendra Krishna Dev said: 'Never was such a stupendous work undertaken with so little preparation or executed with such scanty materials. . . . They have proposed changes in the substantive law and the law of procedure which would not only be most unjustifiable in their effects on the rights of property and would also create constant

dissension.' *Progs. of the Quarterly General Meeting of the British Indian Association*, 1 October 1880.

172. *Bengalee*, 17 March 1883.
173. *Speeches of the Native Members of the Governor General's Legislative Council on the Bengal Tenancy Bill*, p. 16.
174. Ibid., Appendix, pp. I-II.
175. *Som Prakash*, 11 Agrahayan 1293 B.S. (RNP).
176. *The Supplementary Gazette of India*, 21 April 1883, p. 867.
177. *Speeches of Native Members*, op. cit., pp. 20-34.
178. 5 March 1884 (RNP).
179. 11 August 1883 (RNP).
180. 10 March 1884 (RNP).
181. 20 March 1884 (RNP).
182. *The Gazette of India*, 29 March 1884, p. 85.
183. *Reis and Rayyet*, 17 March 1883.
184. *Bengalee,* 14 March 1885.
185. *Reis and Rayyet,* 17 March 1885; *The Gazette of India*, April 1884, p. 277.
186. *Bengalee*, 14 March 1885; *Ananda Bazar Patrika*, 15 September 1884; *Bharat Mihir*, 5 August 1884 (RNP). Asok Kumar Chatterjee, op. cit.
187. Romesh Chandra Dutt was a high ranking civilian; Bankimchandra and Parbati Charan Roy, the author of *The Rent Question in Bengal*, were associated with the provincial administrative department; Gurudas Banerjee (1844-1918) was a teacher and lawyer; Hem Chandra Banerjee (1838-1903) was a *munsif*, later a lawyer in the Calcutta High Court and an eminent poet; Kalicharan Banerjee (1847-1907) was a teacher and prominent leader of Indian Christian; Lalmohan Ghosh (1849-1909), Manmohan Ghosh (1844-96), Umeshchandra Banerjee (1844-1906) and Anandamohan Bose (1845-1911) were barristers; Chandranath Bose (1845-1911) was a teacher and later Deputy Magistrate; Rajnarayan Bose (1826-99) was a teacher; Bholanath Chunder (1822-1910) was a merchant and author; Umeshchandra Dutt (1846-1907) was a teacher and principal of City College, editor of *Bama Bodhini* and a social reformer; Dwarkanath Ganguli (1845-98) was a teacher and social reformer; Rajkrishna Mukherjee (1845-1886) was a teacher and lawyer and; Narendranath Sen was an attorney of Calcutta High Court and editor of *Indian Mirror.*
188. Syed Ameer Ali (1849-1928) was born in Chinsurah, Hooghly. A lawyer and historian, he did M.A., B.L. and became barrister from London in 1873. He taught for a time at the Hindu College and the law department of Calcutta University. He was Presidency Magistrate and a judge of the Calcutta High Court (1890-1904). In 1877, he established the National Muhammedan Association. In 1909, he became

the first Indian to become member of British Privy Council. Books authored by him in English include *The Spirit of Islam* (1891), *A Short History of the Saracens* (1899), *Personal Law of Muhammedans; Muhammedan Law* (2 vols.), *The Code of Civil Practice* and *The Ethics of Islam.*

189. Ameer Ali thundered in the Council:
'I do not propose to enter here into an examination of that somewhat obstruse question—given to the necessity for legislation to regulate the relation of landlords and tenants in the country . . . whether the state by ensuring the zamindars against enhancement and variation of its own demand (and that in fact is the meaning of the Permanent Settlement) had abdicated in perpetuity its legislative function to protect and safeguard the interest of another class . . . a much larger and more permanent class. If the contention of the landlords on this head is correct the result necessarily follows that the government of this country is an incomplete government; that it has in fact established an Imperium Imperio, and that, so far as the ryots are concerned it has delegated all its powers to the ever-shifting body of zamindars.' *Englishman*, 22 January 1879.
190. *Bengalee*, 20 August 1881.
191. *Som Prakash*, 27 Pous 1287 B.S., no. 9 (RNP).
192. Jogesh Chandra Bagal, *History of the Indian Association 1876-1951*, Bharati Library, Calcutta, 1953.
193. Ibid., pp. III-IV.
194. *Bengalee*, 2 July 1881.
195. Ibid., 20 August 1881.
196. Ibid., 2 January 1884.
197. Ibid., 28 February 1885.
198. *Sanjivani*, 5 April 1884. Dwarkanath Ganguli edited this journal. Others associated with it include Herambachandra Moitra, Krishnakumar Mitra, Kalishankar Sukul and Pareshnath Som.
199. *Som Prakash*, 28 April 1884 (RNP).
200. Ibid., 7 July 1884 (RNP).
201. *The Supplementary Gazette of India*, 14 February 1885, p. 17.
202. *Bengal Legislative Council Progs.*, vol. XXX, no. 2, 22 August 1928, p. 456.
203. *The Gazette of India*, April 1884, p. 649.
204. Ameer Ali asked:
'Is it fair or reasonable to constitute a rise in the price of staple food crops as a ground of enhancement when hundred other circumstances like the increasing cost of production, increasing cost of the basic necessities of life . . . tend to show the ryot of today in a majority of instances is not a whit better off than the ryot of twenty years ago?' *The Supplementary Gazette of India*, op. cit., p. 17.

205. *Shakti*, 14 August 1884 (RNP); *Sambad Bahika*, 7 August 1884 (RNP).
206. *Bengalee*, 14 March 1885.
207. The Indian Association asked:
'Will the man who holds 500 bighas under a zamindar, and cultivating 40 or 50, sublets the rest, be considered a ryot? As he actually cultivates 40 or 50 bighas, he must be a ryot, and therefore the actual cultivation of the remaining 450 bighas under him will be classed as under-ryot and will be incapable of acquiring rights of occupancy. There again would be lamentable result, substantial middlemen would be protected, but the cultivators of the soil would not be benefited.' *Bengalee*, 5 April 1884.
208. Ibid., 12 April 1884.
209. Ibid., 28 February 1885.
210. Ibid.
211. Ibid., 14 March 1885.
212. R.C. Dutt, 'Presidential Speech', 27 December 1899, Lucknow Congress, in his *Speeches and Papers on Indian Questions*, 1897 to 1900, p. 125.
213. Dutt, in ibid., 1901-2, pp. 4-5.

CHAPTER 4

The Ploughmen's Poet

Amra chash kari anande!
mathe mathe bela kate sakal hote sandhye!!
Roudra othe, brishti pare, banser bone pata nare,
Batash othe bhore bhore chasha matir gandhe!!
Sabuj praner ganer lekha rekhai rekhai dai re dekha,
Mate re kon tarun kab nrityadadul chhande!
Dhaner shishe pulak chnote—sakal dhara heshe othe
Aghraneri sonar rode, purnimari chandre!!

RABINDRANATH TAGORE[1]

It was Rabindranath Tagore, Bengal's greatest intellectual and best-loved poet, who put to music the world of the peasant. The rhythms of their daily lives seemed to come alive with a voice of its own in his poetry. He was indeed a ploughman's poet who gave glimpses into aspects of the humdrum existence of the peasants that was hitherto unexplored. As his tunes seeped into the drawing rooms of the literati, the trials and tragedies of peasants' lives became entrenched in the educated mind. The gulf between the educated urban and the unsophisticated rustic seemed to bridge as all hummed, often unconsciously:

Oh come lets' reap the harvest—
Reap the harvest, reap the harvest.
Land is our boon companion, today at its call
Our courtyard shall remain filled day and night
We shall receive its gift, so we reap the corn.
So we sing—so we work in pleasure.[2]

Tagore made seminal contribution to the shifting paradigms of thought on the rural world in colonial Bengal. His understanding of the actualities of village life went far beyond the literati's prolonged obsession with legal uncertainties in tenurial position of different categories of ryots under a system of administration that was structured on the principle of maximization of revenue. Saddened by the grim realities he was exposed to, during his sojourn in the zamindari estates held by his family, he focused on a programme of rural resuscitation, which he himself named 'rural reconstruction work', that was worked out over a period of time.[3]

THE SPIRIT OF THE AGE

1890 constitutes a significant landmark in the evolution of Tagorean philosophy. It witnessed the initiation of an enduring relationship between the young poet and the rural world as he was entrusted with greater charge of the family's landed property by his father Debendranath who had great faith in the business acumen of his youngest son.[4] In the next half century of his life, rural Bengal was never away from his thought even when he has in far-off Europe or America receiving accolades from the world over and meeting philosophers, litterateurs, artists and political leaders. As his senses were increasingly motivated by the external beauty of rivers, acres and acres of greenery, thatched roofs, bamboo groves, and seasonal changes the poet in him penned eloquently:

> I bow, bow, bow to thee, O my beauty, the motherland Bengal
> The Ganges' bank, the gentle breeze; you have refreshed my life.
> The unbounded field; the forehead of the sky kisses the dust of your feet—
> The little villages are thickly shaded abodes of peace.
> The mango orchards with densely clustered leaves is the playhouse of the Herds-boy
> The still dark water of the unfathomable lake is your night-cool affection
> Her heart filled with honey, the housewife of Bengal carries water home,
> As I feel restlessly eager to address her as 'mother' my eyes fill with tears.[5]

Here the poet emphatically highlights the physical beauty of rural Bengal that so mesmerized him. On the other hand, close proximity to the villages and an intimate understanding of the basics of life and nature, acquired while living mostly in his boat on the Padma ever since he first came to Shilaidaha as a young zamindar, helped the poet to read the inner story of the rural populace, their troubles and daily rigours.[6] Explaining his varied experiences at Shilaidaha and Patisar he wrote that as the tenants in his estate came to him with tales of their 'joy and sorrow, complaints and requests', the 'village discovered itself to me'.[7]

Reminiscing about his father's love for the countryside, the poet's son Rathindranath wrote.

But there is no doubt that his [the poet's] first and deepest love was for the country of melon green fields with their clusters of bamboo shoots swaying gently in the south breeze and hiding villages in their midst, of majestic rivers with their stretches of gleaming white sand—the haunts of myriads of wild ducks, as well as of homely rivulets with sweet—sounding names, meandering in and out through peaceful villages hugging their banks.[8]

As the poet's association with the rural world increased, so did his knowledge of it. He observed:

I was anxious to see village life in the minutest detail. My duties took me to distant parts by rivers, canal and waterways, and here was a chance to see the panorama of life. The everyday tasks of village folk and the varied cycle of their work filled me with wonder. Bred in the city, I stepped right into the heart of rural charm and filled myself with it. Then, slowly, the poverty and misery of the people grew vivid before my eyes and I began to wish that I could do something for them.[10]

As he probed deeper, a new realization dawned about the discomforts of rural life and the progressive deterioration in the quality of life in Indian villages. He was moved to tears by the sad plight of his tenants and common cultivators.

As the impressionable young poet toured his paternal acres, watching life uninterrupted by the pressures of an urban existence, his keen mind stored information that would give a new meaning to his future work.[11] He revealed his inner thoughts, 'I was struck with shame that I was a zamindar, impelled by the money motive, absorbed in revenue returns. Since that realization I awoke to the task of trying to stir the minds of the people, make them shoulder their own responsibilities.'[12]

The helplessness of the villagers during adversities, their inability to battle diseases, the poverty of the common people and their exploitation by all and sundry shocked and angered him. He wrote repeatedly that he was convinced of the need to improve the lot of the peasants.[13] He himself was filled with the urge to make a difference in their lives.[14]

It was this urge that prompted him to address certain pertinent issues related to peasant and rural life in general and accounted for his mental move towards and appreciation of the needs of the countryside. The work that he undertook subsequently represents the clear articulation of Tagore's perception of the nature of the colonial impact on rural society. Unlike his contemporaries, he took the peasant question far beyond theoretical debates on rights of various categories of ryots. He visualized it not in isolation but as a component of the entire rural landscape. It was a point of departure for the poet who raised these issues at a time when the peasant had virtually disappeared from the public discourse. Except for the Age of Consent Bill, it was matters of vital political and constitutional politics that pre-occupied the literati. The moderate leaders of the Indian National Congress steered clear of all divisive issues. Landowners contributed significantly to its fund and were present in large numbers in the Congress sessions between 1892 and 1902. Even among the professionals, viz., lawyers many were either related to erstwhile landlord families or had landed interests.[15] The social background of the moderate political leaders, associated with the Congress since its inception, made it impossible for it to adopt a 'logical stand on peasant questions'.[16] In fact, as the Congress depended greatly on the British Indian Association of the landowners in the early years of its birth,[17] it often adopted contradictory approaches towards issues relevant to the interests of the peasantry in Bengal and elsewhere.

After the enactment of the Bengal Tenancy Act in 1885, interest of the intelligentsia in agrarian issues waned in Bengal as attention was focused upon the Congress and constitutional politics. There was a conscious attempt to uphold the united fabric of Indian society as the nationalist movement gained ground. Congress leaders could not afford to alienate the powerful landed aristocrats at such a crucial stage in the country's political development. Apart from this, the Act satisfied for the time being certain sections of the peasantry and their patrons among the intelligentsia. The limitations would not become visible before the passage of a couple of years. In the meanwhile, whenever the Congress leadership was forced to take a stand on specific agrarian issues they displayed a pro-landlord bias. Even while aware of the shortcomings of the Permanent Settlement, an issue that had been in public discourse in decades preceding the Act, they demanded extension of the Permanent Settlement without pro-ryot modifications to satisfy the zamindars. They opposed cadastral survey in 1893-4 even though it was aimed to protect the peasants from the manipulations of zamindars.[18]

It was R.C. Dutt, representing a small group of pro-ryot sympathetic within the Congress, who at times in the 1890s deviated from this line of approach. Over the years many of his views changed but his years of practical experience of India's agrarian system, of India's poverty and famines gave him an expert's knowledge and the moral authority with which he articulated his policy prescription albeit within the constraints of his time.[19] Dutt's historical analysis of British imperialism needs to be understood for contextualization of his views on its impact on rural society.[20] He ascribed the adverse economic consequences of Brutish rule, which could not be concealed by the 'progressive' face of imperialism to the blunders of economic administration. While the 'Age of Imperialism' witnessed a civilized administration, construction of railways and a vast expansion of the cultivation, recurrent famines exposed the frailty of the agrarian economy and caused immense hardships. In his later works, which kept alive the interest in economic issues even as political aspirations of the literati heightened, Dutt addressed at length the issues of famine and poverty that were so vital to his understanding of agrarian society. He asserted emphatically that 'the poverty of the Indian people was unparalleled in any civilized country' and even by any moderate calculation, successive famines[21] had carried off 15 million people—a population that was 'equal to half that of England'. He rejected as myths the pressure of population, the improvidence of Indian farmers and exactions of moneylenders as probable causes of famines. He highlighted the narrow subsistence of the economy, viz., agriculture and the inordinate demand made upon it for running a top-heavy, belligerent government.[22] Increasing government expenditure imposed a heavy burden on a people who depended for subsistence only on land and who could not save in years of good rainfall and were resource-less in bad years. Agriculture, which was 'virtually the only source of national wealth in India', was burdened by an excessive land tax. Dutt estimated that in Bengal, it was fixed at 90 per cent of the rental and in northern India it was above 90 per cent between 1793 and 1882. The unprecedented increase in India's public debt and home charges that accounted for an annual drain of one-half of the net revenues of the country, estimated at 44 million sterling pounds, worsened the situation.[23]

Dutt explained recurrent families in the country in terms of poverty of the people, which were indicated by low per capita income. His administrative experience in 1876 informed his argument that high land tax accounted for much of the poverty and resourcelessness of the peasants. A cyclone and a tidal wave struck Bengal in that year.

Entrusted with relief work he found in his district peasants buying in years of distress 'shiploads of rice out of their savings' in the form of silver ornaments.[24] On the contrary, elsewhere, viz., western Bengal where rents were higher in proportion to produce compared to eastern Bengal, Madras and Bombay where land tax were much higher, the people were less resourceful and famines more frequent and fatal.[25] Dutt reiterated what he had said earlier that it was the rigidity of the government revenue demand that tightened the moneylender's hold in the economy. Unlike his earlier contemporaries, Dutt represented a class of intellectuals who treated the peasant question not in isolation but as a component of the overall economic scenario. Their vision was not confined to the region of their birth and later activities. Dutt in his later life thereby treated issues pertinent to the needs of the peasantry in their entirety and his remedial prescription were not just for the welfare of the Bengal peasantry.

Seeking the fundamental causes of rural poverty Dutt focused on the increasing burden on land because of the diversion of former artisans and handicraft workers to agriculture; the increasing expenditure of the government and; the escalating land revenue burdens in the ryotwari areas. His remedial prescriptions were thus based on the major premise that the burdens on land should be reduced.[26] He offered some proposals for land reforms: Where the state received land revenue through landlords and the revenue was not permanently settled, he advised that the Saharanpur Rules limiting state demand to one-half of the rental may be universally applied.[27] Where the state received land revenue direct from cultivators, he advised that the rate may not exceed one-fifth of the gross produce of the soil in any case, and that the average of district including dry and wet lands, be limited to one-tenth of the gross produce which is approximately the revenue in northern India. Where the state received revenue direct from cultivators his advice was that the rule laid down by Lord Ripon, of permitting no enhancements at recurring settlements, except on the grounds of an increase of prices, be universally applied. Where the land revenue was not permanently settled, he wanted that the settlements be made oftener than once in thirty years, which is the general rule in northern India and Bombay. He urged that no cesses, in addition to the land revenue, be imposed on the land, except for purposes directly benefiting the land; and that the total of such cesses may not exceed 6¼ per cent (one *anna* in the rupee), in any province of India. He urged that, in the case of any difference between cultivators and settlement officers in the matter of assessment, an appeal be allowed to an independent tribunal not concerned with the fixing and levying of the land tax.[28] Explaining the need for land reforms Dutt observed that famines were caused primarily due to over assessment of land revenue which left the landlords and cultivators without a sufficient margin to withstand the occasional bad harvest. Dutt was emphatic that the ryots should know clearly what the state demanded and what they were entitled to keep as uncertainty paralyzed agriculture.[29]

A joint memorial drafted by Mr. Puckle, an experienced revenue officer, and accepted with minor modifications was dispatched to the Secretary of State for India. It was then forwarded, for consideration, to the government of India and became the subject of a lengthy resolution on the land revenue problems and policies of the government. The suggestions of the memorialists were: half net produce rule for cultivators paying land-tax direct; half-rental rule for landlords paying the land tax; thirty years' settlement rule; limitation of enhancements from cultivators on the ground of increase in prices; limitation of cesses to 10 per cent of the land revenue.[30]

The essence of these proposals was accepted in Curzon's resolution which stated 'It cannot but be their (government's) desire that assessments should be equitable in character and moderate incidence, and there should be left to the proprietor or to the cultivator of the soil—as the case may be—that margin of profit that will enable him to live in ordinary seasons and to meet the strain of exceptional misfortune.'[31]

R.C. Dutt was satisfied that Curzon had accepted in principle that government revenue should generally be limited to one-half of the actual rental in temporarily settled estates. He also hoped that the rule of thirty years' settlement would be extended to the Punjab and the Central Provinces and local cesses reduced. However, the government's silence on the question of restricting its revenue demand in ryotwari areas left him dissatisfied.[32] As regards the 'half-rental' rule for the landlords, while the government was willing to accept it in principle it refused to adopt any fixed quantitative equation. Consequently, the problem of uncertainty of demand persisted in non-permanently settled areas. Curzon's resolution adopted the same principle with regard to the half net produce rule for the cultivators. But it refused to accept the suggestion that the land tax where estimated at one-half of the net produce should not exceed the maximum of one-fifth of the gross produce on the plea that if the gross produce standard was applied generally then there would be an increase of assessment all around.

As to the grounds for enhancements of land revenue demand, Dutt and the memorialists held up Bengal as the model. The Rent Act[33] allowed enhancement of the tenant's due by private landlords only on the ground of increase in the value of land in consequence of improvements in irrigation works carried out at state expense or on account of a rise in the value of the produce based on the average prices of thirty years preceding such revision. The memorialists suggested that the same rule be applied in areas where the state was landlord[34] but Curzon in his resolution, rejected this demand on the ground that to deny the right of the state to a share in any increase in values, except those which could be inferred from the general table of price statistics which in itself was a most fallacious and partial test would be to surrender to a number of individuals an increment which they themselves had not earned. Dutt agreed that rise of prices was a satisfactory indicator of the rising land values but argued that all the real advantages which the cultivator secured from new roads or railway lines were shown in a rise of prices and so could be regarded as the basis of revenue enhancement.[35] Dutt, however, warned that increasing land values generally indicated agricultural depression and in such cases the real beneficiaries were traders and moneylenders, rather than the peasants. In most cases, where the peasant sold his produce in advance or immediately after harvests he never received a fair price.

The government objected strongly to Dutt's linking of famines to over-assessment of land tax in the ryotwari areas. Ignoring the fact that past savings, which could have been much higher if the revenue demand had been lower, shielded against future wants, the former emphasized on the all round failure of crops as the actual cause of famine like conditions. Dutt's claim that it was essential for the peasants to have prior knowledge about the exact proportion of the produce to be left with them after paying governmental dues was ignored. In spite of being forced into a difficult situation within the Congress for his opposition to proposals for certain pro-zamindar amendments to the Bengal Tenancy Act in 1898, Dutt raised certain pertinent issues that the political leadership was not willing to confront. Arguing from the point of an economist, he was vocal in his contention that throughout his period agriculture had stagnated due to factors such as high revenue demand, escalating rural tension, pauperization of ryots, reindustrialization, insecurity of tenure, and population pressure on land and absence of productive investments.[36] Ryots and erstwhile rural industrial classes had fallen back on land as the only means of sustenance. Apart from moderation of land tax, the extension of irrigation works, i.e. canals, tanks and wells was indispensable for agricultural development. Though he did not advocate scientification of agricultural techniques he was critical of the fact that the issue of irrigation had been neglected and public revenues had been diverted in developing other infrastructural benefits. He pointed to the fact that in a period when 225 million sterling had been spent on railways, only 25 million had been invested in extending irrigational facilities. While in Bengal, where rainfall was abundant and the fields were often inundated by rivers, shallow ponds excavated in the fields would suffice, elsewhere canals would help in adding stability.[37]

As far as the system of land management was concerned, it became clear by the close of the century that his views on the Permanent Settlement had undergone a transformation. In spite of his attempts to find a solution to some of the maladies afflicting rural society, his policy prescriptions were in tune with that of the majority of his contemporaries, who represented the literati. It is strange that in spite of his long experience in district administration and his earlier denouncement of the Permanent Settlement, he shifted his stand to the extent of admitting that it had secured the economic interests of the common people and made possible the extension of cultivation. It had created both 'a thoroughly loyal class of landlords and a prosperous class of peasantry in Bengal'.[38] In fact, he remarked strangely:

> Since 1793 there has never been a famine in permanently settled Bengal which has caused any serious loss of life. In other parts of India, where the land tax is still uncertain and excessive, it takes away all motives for agricultural improvements and prevents saving, and famines have been attended with the deaths of hundreds of thousands, and sometimes of millions. If the prosperity and happiness of a nation be the criterion of wisdom and success, Lord Cornwallis's Permanent Settlement of 1793, is the wisest and most successful measure which the British nation has ever adopted in India.[39]

He went even further to state that:

> The Permanent Settlement of Bengal has proved a blessing, not merely to landlords with whom it was concluded, but to all classes of the community.

It has benefited all trades and professions by leaving more money in the country; promoted the well being of various degrees of tenure-holders under the landlords; moderated the rents paid by actual cultivators; and prevented the worst effects of famines. . . .[40]

Dutt even advocated the extension of the Permanent Settlement to other regions. It is significant that his emphasis here was on the permanent fixation of government revenue demand. He realized that appropriation of the increasing proportion of the rural surplus had impoverished the peasantry.[41]

It is imperative to ponder over Dutt's retreat from his earlier critique published when he was at the height of his official career. By now, he had in fact retired from government service. In the intervening period, peasants had revolted in Pabna and its neighbouring regions in 1873 and the government had found it necessary to enact the Bengal Tenancy Act. It is difficult to accept Dutt's contention that after the labours of a century the British administrators had solved in a satisfactory manner the great 'land question' in Bengal, firstly by extending protection to the zamindars by the regulation of 1793 and secondly by extending protection to actual cultivators by the Acts of 1859 and 1885. In 1899 Dutt justified the latter on the ground that it saved the cultivators from undue enhancements. Consequently, the average rent paid by cultivators to landlords did not exceed one-sixth of the gross produce in any district and fell far short of it in eastern districts.[42] This Act, he said, gave to cultivators the much-needed relief from unjust enhancements and eviction by landlords.[43] This made cultivators in Bengal more prosperous, more self-relying, and safer against famines than elsewhere in the country. It is difficult to agree with Dutt here. By the close of the century is obvious that there were different sets of rights to land and 'ryot' did not automatically signify one who cultivated the land. There was growing realization that it was necessary to protect not only the statutory tenants but also the actual cultivators.[44] Nor was it possible to prevent subinfeudation.[45] Besides, no settlement with the ryots could be meaningful where the exploited could become exploiters of their subordinates.

These developments seemed to have escaped Dutt. The defects of the Act of 1885 in fact, escaped many of his contemporaries immediately. Bankimchandra's *Bangadesher Krishak* was published in 1892 in the form of a book. In its introduction, he wrote that his earlier description of the conditions of the peasantry did not hold true any longer. The new Act had curbed the powers of zamindars and secured the rights of tenants. But even though he welcomed it, he did not indicate the extent to which it had been able to transform the existing framework of the land system. He did not express any concern over the fact that the latter was inculcating separatist feelings in the province.[46] Though sympathetic towards the peasantry he was, even as late as 1892, in favour of keeping the Permanent Settlement intact. The Act of 1885 kept intact the rights of intermediary tenure holders. Bankimchandra was silent on how in such a situation; the poverty of the multitude of the peasantry could be ameliorated.

But by the time Dutt's later works were published, the limitations of the Act of 1885 had become visible to all and sundry. Dutt had even opposed pro-zamindar amendments to the Act in 1898. His change of stand may partly be explained by the fact that he was dabbling in Congress politics. The content of his speech in Lucknow Congress may be understood from this angle.[47] He was weighed by the onerous task of conforming to the Congress policy of avoiding divisive issues as also continuing his presentation of a nationalist critique of British imperialism.[48] In his attempt to defend the Permanent Settlement and the existing social fabric based upon it he belittled its disastrous socio-impact on the peasantry—an issue that he took up so strongly and boldly in his earlier works. Nor did he recognize the need for industrialization at least for the transformation of agriculture itself even though he realized that the problems of agriculture had been accentuated by the decline of indigenous cotton industry.[49]

However, by the closing decade of the century the peasant problem came to be considered as a social-cum-economic question, a question of landlord-tenant relations, in which the landlord was privileged to exploit and the tenant obliged to suffer. But land reforms were inextricably tangled with politics. It was this recognition that silenced for the time being demands raised in the preceding decades by a section of the intelligentsia for a radical solution of the peasant question. The Bengal Tenancy Act of 1885 satisfied the protagonists of both landlord and peasant rights among the intelligentsia as its limitations were not immediately visible to most. Besides, political considerations made it difficult to adopt an impartial approach towards the land question. What escaped their attention was that in a region where landed interests were sharply divided on lines of religion the inability to locate a viable solution to the land question could, foster communal tendencies and agrarian issues could be incorporated within the agenda of organized politics. The advocates of complete overhauling of the land system and abolition of Permanent Settlement, viz., *Som Prakash* and Surendranath Banerjee were silenced by developments initiated in 1885. It was only in representations in some prose and fiction and in the rural reconstruction work started by Rabindranath in his ancestral estates at Shilaidaha that the problems of the rural population continued to be articulated.

Meer Masarraf Hossain highlighted in 1890 the exploitation of the indigo peasants in his *Udasin Pathiker Maner Katha* (Thoughts of a Detached Traveller).[50] Though it did not make much of a stir or have an impact on the public mind similar to Dinabandhu's play published three decades earlier, its language, style and content signified a coming of age of the Muslim intelligentsia in the domain of literary attainments. Hossain's novel is highly informative. It attempts a realistic pen-portrait of the internal workings of two local zamindaries—the estates of Mirsaheb of Saonta on the western side of the river Gouri and that of Pyarisundari of Sundarpur. He raises issues vital to rural social life as also Hindu-Muslim relations in the locality. On the one hand, he describes the oppression of indigo ryots by European planters', viz., T.I. Kenny in Kustia.[51] On the other, he highlights the communal harmony that existed in the countryside in the closing years of the century. He refers to the fact that zamindars, both Hindu and Muslim, preferred to employ Hindus as estate officials. Pyarisundari's chief administrative official was Ramlochan and Mirsaheb's was Debiprasad. Hossain refers to internal feuds within the zamindari families and the manner in which estate officials took advantage of these opportunities. It was at the connivance of Debiprasad that Mirsaheb was deprived of his estate by his nephew Shah Golam. This was a malaise that afflicted most landed families in the nineteenth century. What is, however, interesting is that, in spite of the developments of the preceding years, a feeling of empathy still characterized zamindar-tenant relations. Some of the moving passages of the novel describe Pyarisundari's concern for the hapless peasants so cruelly oppressed by Kenny. She bemoaned the chaos in the countryside, seizure of peasant's lands and forcible plantation of indigo. At times even the latter's homesteads were not spared. The novel reveals the terror unleashed by the planters and how universally they were hated. Significantly, in a meeting held at Shah Golam's house, to formulate a plan of action against the atrocities of Kenny, attended by inhabitants of adjacent localities a Hindu zamindar played the lead role.

Hossain, however, also focuses attention on the dichotomous attitude of zamindars in their dealings with planters and their fellow compatriots. While sometimes friendly with the former, zamindars often maintained cordial relations with the latter. That was why, even though Mirsaheb's friendship with Kenny forced him to help the latter, it was with regret that he received the news of his exploitation of Pyarisundari's tenants.[53] That the peasant issue had not yet become an agenda of communal propaganda in the 1890s is clear from the fact that, in his critique of Bengal zamindars he was equally vehement in his denouncement of the exploitative character of both Hindu and Muslim zamindars. Instances of their cruelty on hapless ryots and their womenfolk abound in his literary expositions. He produced 'art with a purpose'. But like most of his contemporaries he failed to realize the significance of the existing land structure as a system of rural power relation. He preferred to distinguish between 'good' and 'bad' zamindars rather than demand abolition of the system.

By this time Hossain had established himself as a litterateur of some repute and his works were widely read. His handling of subjects pertinent to contemporary society informed the thought processes of many of his co-religionists who at this time began to take an interest in the peasantry. The Muslim literati were particularly perturbed by their miserable fate; the majority of whom they realized were Muslims. This awareness was expressed in the corpus of literary outputs produced by them in the last years of the century. Poverty, apathy of social superiors towards the peasants who were most numerous, moneylending exploitation and the immoral acts of zamindars were the themes that were heavily focused upon, depicted and explained. Noushar Ali Khan Yusufzahi in *Bangiya Mussalman* attributed the misery of Muslim peasantry primarily to lack of education.[54] He drew attention to the fact that even in areas where there had been a shift from food-cropping to commercial cropping and the prices received had trebled, living standards had often deteriorated. Enquiries revealed that zamindari and moneylending oppression had increased manifold. Yet, peasants were unable to break away from the shackles because of lack of education, which made them incapable of attempting any improvement.[55] So he made an earnest appeal to his educated co-religionists to quench what he called the peasants' thirst for education. Muhammad Mashenullah in the novel *Budir Suta* welcomed the possibility of the government's attempt to introduce free primary education in the country on the ground that it would immensely benefit the peasants.[56]

It is, however, doubtful whether there was 'a thirst for education' and knowledge among the Muslim peasants in general. The poorer classes in general were apathetic to any education, whether it was the vernacular variety or the English system. There is no real evidence to suggest that these classes had ever been keen on any formal education.[57] The Hindu cultivating classes, viz., the namasudras, pods and rajbanshis could not afford any education, neither could the Muslim peasants. The Director of Public Instruction in Bengal expressed his doubts in 1871, whether there was 'much difference regarding education between the two great sections into which the bodies of cultivators have for many generations been divided'.[58]

The Muslim literati were vehement in their attack on moneylending operations. They lucidly described how the high rental demand forced peasants to borrow regularly thereby, drawing them increasingly into the debt trap from which they could never break away. Moneylenders very often also duped the unsuspecting illiterate peasants into paying continuously even after their loans were repaid. Access to elementary education, it was generally felt, could have made have made a change in their plight.

Another significant issue that had become a recurrent theme in their expositions that would later be exploited for sowing religious discord in the countryside was the nefarious practices of Hindu *amlahs* that ruined many Muslim landlords, who often morally lax, greatly depended upon them.[59] The *naibs*, *dewans*, *mohurees*, *khazanchies*, *sheristadars*, *peshkars* and managers were usually Hindus who controlled the zamindaries.[60] Muslim zamindars were compared to pension-holding puppets that often did not have a correct estimate of the actual income from the estate or the total number of men employed by him. Dismissal of powerful officials was not easy.[61] It could lead to open revolt or even sale of portions of the estate for revenue arrears. Consequently, it was bemoaned, that over the years the number of Muslim zamindars in Bengal had dwindled significantly. The treachery and cunning of Hindu servants and the love for luxury and lack of foresight of Muslim zamindars were blamed for such a situation. The Muslim literati advised the latter to regularly survey their estates, keep account of annual incomes and expenditures, reduce expenses, supplement incomes, deal strongly with employees; become their own salaried managers and cover their own household expenses from their salaries and refrain from unnecessary and prolonged legal suits. In dire necessity, they were advised to employ suitable, honest, educated Muslims as managers and pay them well.[62]

Thus the signs of a communal overtone was gradually becoming strong in the Muslim intelligentsia's approach to the peasant question by the close of the nineteenth century. Their sympathy for co-religionists was becoming apparent. The deterioration in the condition of the Muslim masses and the simultaneous improvement in the economic well-being of a substantial number of Hindus struck a discordant note. There were at times even appeals to the government to adopt stringent measures for the protection of the peasants, viz., provide for representatives of peasants in the Legislative Council; ban rent enhancements without prior approval from the former; prevent zamindars from interfering in the survey of the record of rights; management by the former of the estates of recalcitrant and oppressive zamindars who exacted illegal *abwabs* from tenants; direct collection of revenue from tenants and; imposition of a ceiling on the rate of interest that was charged by moneylenders.[63] In fact, the government was held responsible for the tragic fate of the masses. It had erred by not controlling oppressive zamindars. Things had come to such a pass that peasants could not hope to get any reprieve even in law courts.

What is significant is that, almost nowhere in the literary works have these upholders of peasant rights cited instances of powerful resistance put up by the oppressed and the exploited. It is strange that the revolt of the peasantry in 1873 did not make much impact even upon Masarraf Hossain. He had been so inspired by *Neel Darpan* where the powerful resistance by Torap had made it a gripping tale. But none of his later works reflect this influence. In fact, none of them experimented with anti-colonial discourses or demanded a change in the existing land system based on the Permanent Settlement, a source of many an agrarian problem. Even the Muslim associations were silent on this issue. Towards the close of the century, branches of Muhammedan Literary Society of Calcutta[64] and the central National Muhammedan Association of Calcutta[65] were opened in the countryside. Other similar associations, viz., the Dacca Muhammedan Friends Association, Muhammedan Union in Calcutta and Malda Muhammedan Association[66] also started functioning, aspiring for a wide mass base. They highlighted zamindar-tenant relations and the destitution of Muslim peasants but demanded no alteration in the existing framework. They preferred to hold lack of enterprise on the part of Muslim zamindars and the intrigues of their Hindu *amlahs* responsible for the ruin of Muslim estates. There were no suggestions for solving agrarian problems, viz., modernization of techniques, and increase of yield and improvement of agrarian relations. There were only appeals for legal safeguards for protection of peasants and permanent restrictions on the powers of zamindars to evict tenants and enhance rents.[67] Leaders like Abdul Lateef and Syed Ameer Ali in the last decade of nineteenth century sought government help to redress the grievances of peasants. The latter did attempt a rudimentary agrarian programme. He suggested rent reduction, government management of estates of inefficient zamindars, prevention of *abwab* collection, fixation of a maximum rate of interest, easy credit facilities and basic education for ryots. Later leaders like Delawarr Hussain Ahmed Mirza displayed similar attitudes. He wrote essays where he suggested methods by which Muslim society could be reformed. In his famous work *Essays on Muhammedan Social Reform,* he detailed the economic condition of Muslims.[68] He described their gradual impoverishment and the ruin of the erstwhile wealthy section during the course of British domination. Though he did not dwell at length on the land problem, he emphasized the fact that innumerable Muslim zamindars had lost their lands and Hindus, many of whom had prospered through trade and commerce, had bought these up.

The agrarian perceptions of the Muslim intelligentsia over the years were entangled with their political beliefs. Their main concern was Muslim zamindars and peasants. Where their concern was the former they on the whole blamed Hindu estate officials. The oppressors of Muslim peasants, except in the early works of Masarraf Hossain, were usually Hindu zamindars and moneylenders. They were silent on the fates of multitudes of Hindu peasants whose condition was no better. Nor were Muslim zamindars and *kutcherry* officials less oppressive. The Muslim literati did not propose a comprehensive plan of rural reforms.

NEED FOR RURAL REFORMS

Tagore realized very early in life that the problems of the peasantry could not be solved in isolation. Prolonged stay in the countryside awakened in him an urge for rural upliftment, quite a novel idea at the time. He delved deeply into the socio-economic problems of the villages. His economic thinking focused on the countryside. A humanistic approach informed his ideas.[69] The central problem that attracted his attention was the general decay he witnessed around him. An idealist, a thinker, a poet and a devoted worker then joined hands and started on an identical mission, the objective of which was rural resuscitation. Away from an active political life, he devoted himself to the task of revival of villages, introduction of welfare measures, developing agriculture and cottage industries and promoting self-reliance. The principle of cooperation underlay all his activities. Songs, poems, essays, travel diaries, letters, novels and short stories produced by him give us an insight into a thoughtful mind. Unlike any of his predecessors or even later contemporaries, he attempted to give practical application to his village-centric ideas, formulated through acquaintance with the day-to-day lives of the common people, in his zamindari estates at Shilaidaha, Patisar and Sriniketan. He constantly traversed from the realm of ideas to the realm of action.

In the poet's understanding the village constituted the kernel of the entire socio-economic structure of the country. It was the nucleus of Indian civilization. He held degeneration of villages responsible for all maladies afflicting the country. Superstitions, lack of education and knowledge, inability to cooperate with one another accounted for rural decay.[70] Explaining his varied experiences he wrote:

> I was anxious to see village life in the minutest detail. My duties took me to distant parts by river, canal and waterways, and here was a chance to see the panorama of life. The everyday tasks of village folk and the varied cycle of their work filled me with wonder. Bred in the city, I stepped right into the heart of rural charm and filled myself with it. Then, slowly, the poverty and misery of the people grew vivid before my eyes and I began to wish that I could do something for them. I was struck with shame that I was a zamindar, impelled by the money motive, absorbed in revenue returns. Since the realization I awoke to the task of trying to stir the minds of the people, make them shoulder their own responsibilities.[71]

He agonized over the helplessness of the villagers when confronted with diseases and natural adversities. His short stories, viz., 'Panchabhut', 'Palligram', 'Shasti' describe his perception of the rural world.[72] The urge to undertake reforms became strong.[73]

The villages, in the poet's perception, had once been thriving and prosperous. But during British rule they had lost their pre-eminent position to fast emerging towns.[74] British rule had ignored the interests of the village and the village system had almost collapsed.[75] What was far more alarming was that it had destroyed all initiative and enterprise in the villagers. Education, adequate food supply and even basic medical facilities were denied to the inhabitants. As the main artery was gradually blocked, blood circulation in the country's society and economy stopped. A process of slow death was thus initiated. Villagers became dependent on the administration for whatever benefits they received and did not attempt to help themselves.[76] Rural development was the need of the hour.[77] He expressed this idea very forcefully during the anti-partition struggle in Bengal in 1905-6.

The poet advised political leaders to channellize political awareness of the masses towards constructive work. He suggested its inclusion in the political agenda of the anti-imperial struggle. Villagers, he believed, should be taught to take responsibility for their own lives.[78] They should be trained in the art of self-reliance. In ancient times it was the duty of the ruler to protect his kingdom, conduct war, maintain peace and dispense justice.[79] Society was responsible for all else. It monitored civil life, provided education, drinking water, medical aid, welfare, etc. Unlike in Europe, Indian society did not depend on the government for basic amenities of life. Through the ages it was civil society which taught people the essence and values of life and helped them to develop their full potentiality, their *atma-shakti*.[80] Colonial rule fatally affected this traditional rural society. Towns prospered while the rural economy was impoverished. Poverty was the bane of rural life. To the impressionable young poet the ryots, helpless as 'new-born' children, were assigned to a life of unending pain. He understood that the situation was further worsened by their apathy and reluctance to change. Poverty was thus not just 'exogenously created' by colonial impact, it 'remained in the social body itself'.[81] The 'fear of poverty' was present.[82] Lack of education had made the common rural folk susceptible to exploitation from all quarters. He bemoaned the fact that the English-educated Indian youth residing in urban areas had failed to realize that without revival of decadent villages, the nerve-centre of Indian civilization, the resurgence of the country would remain a distant dream. But even while appealing to them to join him in his mission he insisted on motivating villagers to promote their own cause. First, they should be inspired to undertake their own resuscitation and then recognize the strength of their combined efforts.

Rabindranath differed from the politics-centric 'nationalism' of his contemporaries. To him the real basis of *swaraj* lay in the development of the full potentiality of the 'self'. Its absence in the villagers contributed to the shortage of adequate food and lack of education, knowledge, health and joy in the country.[83] Previously, he said, a few healthy men used to be the mainstay of the village and its guide. Rural welfare was their responsibility. This undermined the common person's ability for self-reliance.[84] He enumerates:

> In my estate the river was far away and lack of water was a serious problem. I said to my tenants, 'If you dig a well, I shall get it cemented'. They replied, 'You want to fry fish in the oil of the fish itself! If we dig the well you will go to heaven through the accumulated virtue of having provided water to the thirsty, while we shall have done the work'. The idea, obviously, was that an account of all such deeds was kept in heaven and while I, having earned great merit, could go to the seventh heaven, the village people would simply get some water. I had to withdraw my proposal.[85]

To his mind, trapped between the tyranny and charity of the powerful men in the villages, the common folk were deprived of all feelings of self-esteem. Woes of present life were attributed to sins of previous births. They were unshaken in their belief that they could not escape their fates. So, for any fruitful attempt at recovery it was first necessary to awaken their minds through the development of the 'self' or *atma-shakti*. Explaining the need for developing the inner strength, he informs:

> There is a story that a kid went weeping to the Lord and said, 'Lord how is that all creatures seek to devour me?' The Lord replied, 'What can I do, my child? When I look at you, I myself am so tempted.'
>
> Even god cannot guarantee that justice will always be done to the weak and undeserving. Why then we do persist in vainly seeking justice from the administration or the Parliament? Good intentions are irrelevant in this context. Law itself becomes weak in contact with the weak. The police become a terror; he from whom we seek redress against the police becomes its greatest patron.
>
> We must also realize that it is against the present-day policy of our rulers to attempt to strengthen their subjects. The very man who, in the exercise of his judicial functions on the police commission declared the police guilty, in his administrative capacity finds it necessary to protest in tearful tones against the slightest check in their irresponsible power. He is perhaps afraid that if the subjects gain in strength, they may prove too strong for him as well. If the kid be made too tough for other jaws, it may prove less succulent to his palate! Verily are the gods the slayers of the weak![86]

During his sojourn to the countryside, the poet had made two discoveries: first, the villagers seemed to have lost all ability to help themselves and; second, both research and technical assistance would be needed if they were ever to learn to rescue themselves from their creeping decay. Nevertheless, external help could be fruitful only where the person could utilize the help.[87]

PERCEPTION OF RURAL SOCIETY

Rabindranath ascribed the changes in agrarian society during colonial rule to the introduction of Permanent Settlement and subsequent alterations in the system of land tenure. The poet blamed absentee landlordism for destroying the feeling of empathy between the agrarian classes. These landlords drained the rental out of the villages refusing to invest even on basic amenities, viz., new wells, better roads, etc. Those that existed fell into disrepair without proper maintenance. Ryots fared worse when zamindars leased out their lands to intermediary tenure-holders. Incessant increases in rent, extraction of other dues and evictions, unprofitable agriculture, high prices and credit needs of the peasants completed the latter's ruin. Recovery of agrarian relationships, in the poet's understanding, was an essential pre-condition for rural revival.[88] Rural problems could be solved through reforms in agricultural practices and not in land system.

But till 1930 the poet did not prescribe alterations in the tenurial system.[89] Arguing within the existing pattern he professed to focus on the human element and distinguished between 'good' and 'oppressive' zamindars. The picture of a benevolent zamindar immortalized by Romesh Chandra Dutt and Bankimchandra was in his mind. The 'good' ones had at different times, in Tagore's interpretation of the course of history, protected their tenants against European planters, moneylenders and natural calamities. Earlier zamindars had set glorious examples through their generous attitudes, welfare activities and patronage to art and literature. But changes in the social composition had drastically altered the situation.[90] Yet, their social standing, wealth and access to education made them the 'natural leaders' of rural society. The poet reposed great faith in the well intentioned members of his class.[91] Outlining a plan of work he appealed to them to educate the ryots, create an awareness of their rights and potential. Aware of the pitfalls of absentee landlordism, he stressed the need for residence in their estates and advised investment of a portion of the rental on constructive work. However, he warned:

> However spending in charity from a distance will not do any good. Sometimes ago, I learnt that a highly placed police official, not content with causing serious loss to some fishermen, had harassed their entire village under pretence of a local enquiry. I called the fishermen to me and offered to pay for the services of a senior counsel from Calcutta to defend their rights. 'Master!', they said with folded hands, 'what shall it profit us even if we win the law suit? The police will remain with us always and we shall never live in peace again'. I thought within myself that they were right. The weak will suffer even if we win. The only real gift is the gift of strength; all other offerings are vain.[92]

Once awakened the tenants could defend themselves against tyranny from any quarters, be it the tyranny of the landlord or someone else. He said:

> Otherwise the tenants can never be guaranteed their rights through mere legislation or government favour; their weakness will always tempt whoever sees them. How can the manhood of the country be restored if the majority of our people remain easy victims for the landlord and the moneylender, the police and the revenue officer and the clerk of the law court? And if our countrymen do not learn to be men, how can they be taught to be a sovereign people?[93]

The poet's reform proposal included strict vigilance of the activities of estate officials who were often the worst exploiters, particularly in case of absentee landlords.

Through practical demonstration, Rabindranath himself tried to uphold the role of an ideal zamindar. At Shilaidaha, Patisar and then Sriniketan he undertook certain development schemes for the benefit of ryots, viz., construction of roads, wells and canals for irrigation purposes. In Shilaidaha, a cess called *kalyanbritti* and in Patisar *hito-ishibritti* were collected from the tenants to meet a part of the expense incurred. The rest was financed by the estate.[94] Clearly he believed that the interests of the ryots could be best secured with the 'good' zamindars. He reposed the same faith in the class that Cornwallis did when he introduced the Permanent Settlement. He expected them to be actively involved with the production process and adopt a paternalistic attitude towards the ryots.

In his attitude towards the zamindari system, Tagore displayed certain contradictory trends. Income from their estates supported the lifestyle adopted by members of his extended family.[95] Apart from

Satyendranath,[96] the poet's elder brother, none of the others took up independent professions. Immensely talented and creative, the Maharshi's children and other members of the family emerged as leading luminaries of Bengal's cultural pre-eminence since the last quarter of the century.[97] But these pursuits were hardly remunerative. On the contrary, secure income from landed estates sustained many of these pursuits. In the subsequent period as the defects of the Act of 1885 became apparent, the need for amending certain sections became stronger. Among the most hotly debated issues, the proposals for granting the right of free transferability to occupancy ryots and endowing the right of occupancy on under-ryots and *bargadars* were most crucial.[98] Relations between zamindars and tenants reached the nadir. Both classes began to clamour for modifications in sections of the Act. Zamindars held it responsible for the creation of more occupancy holdings, hindering rent enhancements and providing opportunities to ryots to transform the nature of holdings on the pretext of improvement. Litigations vitiated agrarian relations. Landlords also objected to the grant of the right of transferability to occupancy peasants and the right of occupancy to *bargadars*. Ryots on the contrary, vehemently resisted restrictions on their power of transfer of their lands by sale or mortgage as also rent enhancements. They demanded the right to fell trees and build houses within their holdings. Nevertheless, they too were not agreeable to the grant of occupancy rights to *bargadars*.

On these issues, both classes found their advocates among the intelligentsia in the 1920s. One of the earliest to make a statement in eloquent prose was Pramatha Chaudhuri, the son-in-law of Rabindranath's elder brother Satyendranath and an eminent literary personality in his own right.[99] He was for long associated with the poet in the work of estate management that sharpened his insight into the existing agrarian scenario. The bulk of his thoughts on the subject is contained in the elaborate essay 'Rayater Katha'.[100] In a startling deviation he proposed that Bengal ryots should be made owners of their holdings. Land, in his opinion, belonged to those who tilled it. In his interpretation of land rights prior to colonial domination, peasants had been in possession of proprietary right but had lost it since the introduction of the permanent settlement. He now wanted its restoration to them. However, he stopped short of advocating dismantling of the structure built-up over the years. His familial connection with the landed society of Bengal explains this conservative trend. He favoured implementation of legal changes to safeguard ryots within the existing system. What is significant is that he defended his proposal on the ground that it was necessary to push through such modifications to thwart all threats of a social revolution of the Russian type in future.[101] This is an indication of the familiarity of the literati with developments in the world and the fear of possible threats to the established social order. Chaudhuri was categorical in his assertion that if the latter was to be maintained then changes in tune with the changed circumstances of the post-war period were indispensable. Or else there was the possibility of peasants protesting against age-old exploitation and seizing by force what they considered was rightfully theirs. Pondering over such eventualities, Pramatha was actually questioning the moral foundations of the existing social order built upon un-egalitarian principles. The Russian experience informed his intellectual leanings.

In order to prevent a reordering of rural society, Pramatha suggested an alternative plan of limited reforms. He proposed that to enable ryots to become peasant proprietors, the occupancy ryots' right to free transferability of their holdings be legally recognized. In his understanding, it was customarily recognized in some areas. However, where it was not, zamindars extracted *salami* from each such transfer,[102] the amount depending on the extent of their land transferred. Pramatha categorically rejected Rabindranath's suggestion in a rejoinder to 'Rayater Katha', that it would be prudent to deny the right to transfer *jot*s to the ryots on the ground that it would prevent increasing mortgage of the latter's land to moneylenders and consequent landlessness. He was convinced that in spite of the threat of such eventualities, the peasants' right to transfer their holdings should be maintained as it was a proprietary right and should be vested in he who was the actual tiller. On the contrary, zamindars should be denied the right to extract compensation on each such transfer. Mode of production would not be affected if such transfers occurred between ryots. Matters would be complicated if a non-agriculturalist or a *jotedar* acquired a ryot's land. If that happened, the latter would have the land cultivated by an *adhiar* or *korfa* tenant, who would not be entitled to any of the rights that ryots enjoyed under zamindars. Since buying and selling of land could not be prevented, Pramatha suggested that it should be safeguarded through legal restrictions. Where land was transferred to a *jotedar* by a zamindar, the occupancy right of respective ryots should be left undisturbed. He left it to legal experts to formulate necessary policies as he himself was unable to suggest how this could be done.

Chaudhuri's proposal also included the suggestion that the peasant's *jot* should be made *mourasi* and *mokarari*. A ban on all enhancement was desirable.[103] As long as the Permanent Settlement between the state and zamindars existed, a similar contract between the latter and ryots should also be in force.[104] He claimed this demand was not a new one. He recalled that Raja Rammohun Roy way back in 1831 had observed that ryots would perish if their rents were enhanced. So zamindars should no longer enjoy the right to increase the rent of the peasants at will.[105] Rent, in his view was fixed by the record of rights according to the land of the peasant and it should be legally permanent. Regarding other rights of peasants, he was of the view that they should customarily have the right to fell trees in their own *jot*s. Wood was necessary to meet daily requirements; hence they could not be fined by zamindars for chopping down trees which they had probably planted. As to the right of peasants to dig wells and build houses on their lands, Pramatha felt that the law was very confusing. While it recognized the former's right to introduce improvements on their *jot*s it was not very specific about what that entailed. If the ryot, as a case in point, built a *pukka* house on his *jot* at his expense, the zamindar could either fine him or bring an eviction suit against him.

Pramatha blamed the ryots' lack of awareness of their rights and education for their 'state of slavery' and poverty. Access to primary education, the right to transfer land, cut trees, build *pukka* houses and conversion of their *jot*s into *mourasi mokarari* would enable them to become peasant-proprietors in the real sense.[106] Only then would the wealth of the country increase and the race would become, in Pramatha's understanding of the trends in world history, as powerful as was the case in France. All this, he advocated, would be easily possible by minor modifications in certain clauses of the Tenancy Act. Instead of altering the existing framework of land relations, this would on the contrary make good the promises made to peasants by the makers of Permanent Settlement. Pramatha did not suggest an alternative system even though in his reading, its framers misunderstood the traditional land structure and gave to zamindars rights which they never had. Zamindars were just tax-collectors while the ryots had certain rights which were now denied. He now suggested the restoration of these rights to the latter as also the bestowal of occupancy rights to *bargadars*. Neither Act X nor the Bengal Tenancy Act had much improved the position of ryots and were just half-measures.

Chaudhuri did not repose the same faith that the poet had in zamindars. He was not comfortable with the idea of isolating good zamindars from bad and did not regard them as the natural protectors of ryots. In view of Rabindranath's attempt to improve the lot of tenants in his estate, Pramatha considered him a 'unique' zamindar. None could compare with him. The condition of ryots in almost all the estates were pathetic. Consequently, agriculture had languished in Bengal. Development of human resources under the circumstances was imperative.

In exploring Pramatha's approach to pertinent socio-economic problems, it is crucial to review the influence of the First World War and its aftermath and the Bolshevik Revolution. In fact, the bulk of the creative literature produced in the inter-war years was inspired by the European ideal, particularly contemporary French and Russian literature. There was a similar attempt to focus on the common person, describe the daily ordeals, and minutely analyse human psychology and the vicissitudes in political and economic spheres. Rural issues were addressed. The first sparks of socialist thinking is to be found is some of the writings in *Sabujpatra*. But the impact was more pronounced in the creations of the 'Kallol group' led by Dinesh Ranjan Das.[107] Periodicals, viz., *Upasana*, edited by Radha-kamal Mukhopadhyay, *Langal* and *Ganabani* by Muzaffar Ahmed and the poet Nazrul Islam set a new pattern. Socialist thinking was no longer a matter of speculation. It was slowly becoming an ingredient of Bengali literary output. By the 1930s its impact was an acknowledged fact as visible in some of the writings of Manik Bandyopadhyay, Bibhutibhushan Bandyopadhyay, Sailajananda Mukhopadhyay and Gopal Haldar.[108] It reached a high water mark by the time of the Second World War. But its influence on Chaudhuri was not a significant one. In fact, some of his publications in *Sabujpatra* reveal a latent resistance to Marxist ideas. His appreciation of the problems of the peasantry and other socially downtrodden may be explained as a manifestation of a sort of emotional humanism that stemmed from his search for a deeper and sincere understanding of social maladies. Nevertheless, his vision was often blurred. Like his mentor, he too left the basic problems of land relations unaddressed.

He was not averse to empowering ryots with limited rights within the existing framework, as a safeguard against a possible social revolution of the Russian type.

Rabindranath, on the contrary, till much later was not willing to concede to the principle that land belonged to those who tilled it. He defended himself on the ground that if given this prerogative they would be unable to retain it. Land could belong to ryots only if it became saleable property and restrictions were imposed on its transferability. The chances of ryots buying land in an open market, the poet was convinced, were remote and non-agrarian classes, viz., moneylenders and marwari traders would increasingly dominate the land market. In such an eventuality whatever rights ryots enjoyed on zamindari estates would no longer be secure. Under the circumstances, the poet felt it prudent to leave proprietary rights undisturbed until such time when peasants could be entrusted with it. Besides, the law of inheritance would complicate matters by causing greater fragmentation of peasants' land. Soon the small plots would be inadequate to meet the extended needs of peasants' families. In the early 1920s, Tagore opposed the proposal to grant the right of free transferability to occupancy ryots on these grounds. In order to strengthen his argument he cited, the fact that he himself had succeeded in thwarting ryots in his zamindari from selling out to moneylenders by imposing restrictions on land transfers. The picture of benevolent zamindars that emerged as saviours of ryots during the heyday of European indigo planters was too strong in his mind for him to visualize an alternative. He was convinced that restrictions on land transfers had saved ryots and so should remain in force till such time when the latter was better able to enjoy the right to unrestricted freedom of transfer.

However, the poet did see some sense in modifying the existing law in favour of ryots by fixing their rental demand. He was at one with those of his contemporaries who thought it judicial in light of frequent enhancements and litigations that hindered productivity. He also agreed with Pramatha on obliteration of restrictions on peasants' rights to fell trees; build *pukka* residential structures and dig ponds on their land by way of redressing some of the old grievances of erstwhile tillers who resented them. This would also help in fostering cordial production relations. He also assigned to zamindars a paternalistic role in rural society. He held himself up as a role model.

In all the work of rural reconstruction that he undertook, he was motivated by this idea. In fact, at Lahini, he tried to build up an ideal village that was to be an ideal for the rest of the country.

To Rabindranath it was in the village that one could see the real motherland. He wrote:

> We see our real motherland in the village; her heart is there, and it is there that Lakshmi, the Goddess of Plenty, seeks her throne.
>
> For a long time that throne has not been kept ready for her, for the opulent demi-god Kuver has lured men's minds away to the city's *yakshapuri*. We have long neglected to invoke Lakshmi to her own sphere of plenty; thus from the land have vanished health and beauty, knowledge and joy, and even of life itself little remains. To-day in the village the tanks are dry, the air pestilent, the roads impassable, the granaries empty and social bonds lax. Envy and malice, squabbles and misdeeds hasten the decay of the crumbling society. The end seems near, for in this squalid, uncared-for land the fearful rule of Yama grows more powerful every day.[109]

Rural revival, in the poet's understanding, was the prime need of the hour. This could only be possible through cooperative efforts.

THE SPIRIT OF COOPERATION

The spirit of cooperation underlay the poet's entire attempt at rural reconstruction. He wrote:

> May the Giver of all Bounty be pleased with all those who have to-day taken upon themselves the task of filling with milk the drained breasts of the village—that nourisher of life; who are lighting a lamp to bring radiance to her gloomy and joyless home. May these dedicated souls, by their sacrifice, service and devotion, by unifying all in harmonious living and coordinating all disjointed efforts, banish from India the sins of her people, accumulated through centuries of ignorance and inertia, and with these the curse of an angry deity; this is what I pray for with my whole soul.[110]

The moving spirit behind the popularization of the cooperative idea in Bengal, the poet believed that a complete man was one who had this capacity for union. Man lost his true stature when he failed to unite fully with his fellows. A lone individual was a fragmented being. The fear of poverty could be countered if men stood and acted in a group. It was only by combining, the poet asserted, that man had achieved all that was worthwhile in life—knowledge, faith, power and wealth.[111]

When Tagore spoke about cooperation, the idea was a novelty in the country. The cooperative movement in Europe, particularly in

Denmark and Ireland informed his thinking. The cooperative idea evolved gradually in his mind and he attempted to give it multiple applications. 'What is civilization', he asked, 'but a state of union in which the strength of each individual adds power to all and the power of all fortifies the individual?'[112] He blamed the abject poverty of the country on the people for trying to keep them segregated and bearing all their heavy burdens by themselves. But the people of Europe realized that their combined efforts could be their strength and their capital. Tagore held firmly to the belief that the mind of the civilized man had gained in stature by the union of the thoughts of many; likewise, the combined work of many could attain a new magnitude. It was only by the application of the cooperative system that the country could be 'rescued from its age-long poverty and stagnation'.[113] He advocated that it was necessary for leaders to make their compatriots realize that what was not possible for a single individual would be possible when fifty united in a group. Man was by nature gregarious. Singly he was not complete; he was fully himself only in conjunction with others.

As it was man's nature to live and work with others, he observed, his welfare and prosperity lay in fulfilling his nature. When many lived as one then each benefited from the combined strength of the many.[114] There was no limit to the benefits that product of cooperation conferred on humanity in the spheres of religion, finance and education. The welfare of the individual as well as the community was achieved only when there was scope in a community for each member to work for the good of all. It was this principle which had made man great in knowledge and given a moral basis to his conduct of practical affairs. Where it was lacking, there was suffering, malice, falsity, barbarity and strife. Where ignorance or injustice prevented man from working for the common good, the result was misery and misfortune. Addressing the issue of poverty in India the poet observed:

> It is not enough to say there is a shortage of funds in our country; worse there is a shortage of hope. We cast all the blame on fate as we bear torments of hunger. We grovel in the dust, assured that only the mercy of Heaven or of people from outside can save us. It does not strike us that the remedy is in our hands.[115]

To the poet it seemed far better to instill hope in the heart than to offer alms. He who was lacking in hope could only perish. No one could save him by offering alms or some other help.

The poet expressed his sadness at the sight of educated men in the country eager to serve the country, nursing the sick, feeding the hungry and offering alms to the poor. These seemed to him nothing short of attempting in vain to put out a fire by blowing on it, when it had enveloped the whole village. Mere treatment of symptoms was not sufficient to cure the malady. The causes of the disease would have to be removed for an effective cure. For this he suggested that people must at first 'cease to be parochial'.[116] They should feel at one with the outer world. In the economic sphere, their efforts would then have to be coordinated to the efforts of men elsewhere.

The idea of cooperation he stated wistfully was practised in the village economies of ancient India. In those days, as he interpreted, while the ruler discharged duties related to war and peace, administration and justice, it was society that supervised and monitored all else, viz., civil life, provided education, mental and physical nourishment. It was through combined efforts of the locals that common problems were solved. But life, he felt, in contemporary society had become more complex. The masses needed to realize their own strength as villages had decayed. If the Indian economy could adopt the principle of cooperation in all spheres of activities the work of revitalization of villages would be made easier and the country would get a fresh lease of life.

It was in his rural reconstruction work at Shilaidaha, Patisar and Sriniketan that the best application of his idea of cooperation was witnessed. It was not an easy route that he journeyed upon. But repeated failure of his ventures could not shake his belief in the principle.[117] In fact, he was talking of cooperatives at a juncture when his countrymen were unfamiliar with the concept.[118] An understanding of the problems that he was confronted with and an appreciation of the solutions that he indicated would enable one to realize his sincere solicitude for the 'toiling masses of humanity'.[119]

In 1904, the Government of India enacted the Cooperative Credit Societies Act. It was held in official circles that only through cooperative societies could the problem of rural credit be solved. But the cooperative movement failed to take off immediately. Rabindranath, however, was not content with restricting himself to setting up cooperative credit societies alone. He wanted the cooperative principle to envelop all types of activities associated with rural life, viz., agriculture, industry, poultry, dairy farming, etc. Even recreational activities were to be organized on similar lines. This would help make the villages self-sufficient. In 1912, the government passed an Act to enable registration of all types of cooperative societies.

Rabindranath enthusiastically supported all types of cooperative ventures initiated by others. He promoted, for instance, the National (later Hindustan) Insurance Society which was established on 13 February 1909.[120] He even made available a room, in his house in Jorasanko, to be used as its office. In fact, he was one of the signatories to an appeal[121] made by some prominent men of Bengal who had 'agreed to subscribe shares according to our means' for 'similar support and encouragement from others to the scheme of the National (later Hindustan) Cooperative Society.[122] His nephew Surendranath Tagore was one of its most enthusiastic patrons. Recalling the reasons that prompted his association with the society, Rabindranath observed in his Presidential Address at its Silver Jubilee on 13 February 1934 in Calcutta, that he had been motivated by his own strong belief in the principle embodied in its constitution:

> When over twenty-five years ago, the scheme of this Insurance Society was laid before me, a picture of the long and arduous road that needs must be traversed by such an institution, flashed vividly through my mind. But it was this very difficulty of achievement that chiefly attracted me to its programme, and the other attraction was strangeness of the spectacle that it conjured up, of our Bengali countrymen thus bonding together to organize a vast wealth-producing organization on up to date lines.[123]

The cooperative principle was, to the poet, an ideal and not a mere system and therefore it could give rise to innumerable methods of its application. The manhood of man was at length honoured by the enunciation of that principle. It would not lead man into a 'blind alley' as at every step it communed with his spirit. Therefore it seemed to the poet, in its wake would come, not merely food, but the goddess of plenty herself, in whom all kinds of material food were established in an essential moral oneness.

Those of the Irish idealist poet George Russell popularly called A.E. heavily influenced the poet's ideas on co-operation.[124] The latter advocated the need for collective action to solve agrarian problems and firmly believed that poverty, disease, joylessness and backwardness of rural areas could and should be removed by co-operative efforts. It made the poet realize what varied results 'could flow there from, how full the life of man could be made, thereby'. He could thus understand how great the 'concrete truth was in any plane of life, the truth that in separation was bondage, in union liberation'.

The pioneer of Irish agricultural cooperation, that strongly influenced the beginning of the agricultural cooperative movement in Great Britain and the Commonwealth, inspired the poet further in his endeavours. Sir Horace Plunkett's attempts at economic reconstruction of Ireland through cooperatives motivated him primarily because of the similar political conditions in both countries. Plunkett in fact convinced the poet to adopt the Irish model rather than the Danish one.[125] In Denmark, the state had played a vital role in the success of the cooperative movement, viz., in the, case of dairy-farming. The Danish state, unburdened with heavy defence expenditure, invested heavily in welfare schemes. Arrangements were made for the extensive training of the people at large in dairy-farming which was possible only in a 'free' country. He bemoaned:

> But it does not rest with us in India to disburse the revenue for purposes of the health and education of the people. The amount set apart for the country's welfare is hopelessly inadequate for these purposes. Here again the problem is the extreme difference between the powers of the state and the powers of the people. But we must conquer our poverty and downfall due to this difference by realizing our own strength through methods of cooperation, by improving our health and education.[126]

In a country groaning under colonial exploitation, adoption of cooperative methods seemed to him the sole possible solution to the problem of poverty as he observed,

> I feel that the separateness of self-interest which had so long contemptuously ignored the claims of the truth of man, was at length to be replaced by a combination of common interests which would help to uphold that truth, proclaiming that poverty lay in the separation and wealth in the union of man and man. For myself I had never believed that this original truth of man could find its limit in any region of his activity.[127]

The poet's faith in the efficacy of the idea of cooperation was shared by many of his younger contemporaries who were motivated by his attempts to actualize his ideas in his rural resuscitation efforts.

SCIENTIFIC AGRICULTURE

Unlike most of his contemporaries, Rabindranath was convinced that in an agricultural country no social revitalization was possible without agrarian development and thereby attempted to popularize the concept of scientific agriculture. The lack of sustained government initiative in this direction could not have escaped a keen and perceptive mind. The idea of experimental farms was familiar since the establishment of the government's early venture at Saidapet to experiment with machines, implements, manures, crops and livestock.[128] However, the colonial government was unable to adopt a long-term comprehensive agricultural programme. The private agricultural societies were left to upgrade and improve the agricultural sector. The government provided aid 'only when goaded' by the successive Famine Commissions.[129] The poet's departure from the general trend was prompted by the fact that from the 1890s onward certain attempts were made both at the official and unofficial level to rectify the unscientific nature of the prevalent system of agriculture. In 1890 Dr Voelcker was sent to India to report on the existing system of agriculture and suggest reforms.[130] He highlighted the unscientific and backward nature of agrarian practices. It is probable that many of the contemporary educated agricultural enthusiasts were aware of the report and the general scenario. There was a flurry of activities. A lead role was played by Nityagopal Mukherjee. He was member of the Bengal Provincial Civil Service, a gold medalist from the Royal Agricultural College, Cirencester, England and Fellow of the Highland and Agricultural Society, Scotland. He became Professor of Agriculture, Civil Engineering College, Shibpur. He wrote a number of agricultural texts.[131] He was particularly interested in the improvement of sericulture in Bengal.[132] The contents of the texts when made known to the illiterate peasants involved in sericulture could prove immensely useful.[133] The poet was certainly not unaware of these trends.[134] His keen mind easily perceived the glaring defects in the existing system. On the one hand, the increasing population created greater demand for food while on the other, ryots persisted with traditional techniques of production and were rather hostile to new ideas and practices. Population pressure forced continuous cultivation and the elimination of the fallowing practice contributing to progressive deterioration in soil conditions. There was also considerable reduction in land under fodder.[135] This adversely impacted upon animal husbandry. A keen student of agricultural development in the world and desirous of acquainting his country men with improved technology, the poet sent members of his family to Illinois University in America to learn the principles and methods of agricultural science.[136] This he hoped would provide him with well-trained workers for the work that he had in his mind for the future.[137]

Rabindranath voiced his thoughts aloud when he highlighted the basics of agrarian developments in Europe and America and outlined a scheme of reforms for his country.[138] Mechanization and scientific agriculture held out the prospects of better utilization of time, labour and lesser expenses for the future. However, its initial costs being very high, made it a somewhat difficult option for small peasants and holders of tiny *jots*. Introduction of cooperative farming in each *mandali* or village was advised. Tagore mooted the idea of amalgamation of fragmented lands, common use of tractors and collective agriculture. Plough, loom, bullock cart, oil mills, etc., helped to make agriculture, weaving, extraction of oil, sugar making and all such activities productive.[139] He observed that sub-division and fragmentation of land over the years had undermined the productivity of Bengal ryots and suggested consolidation of land holdings and collective agriculture to make scientific agriculture possible and mechanization feasible. Nevertheless, he was aware of the difficulties of initiating peasants to new ideas and practices. They had become so accustomed to their hard lives that they had resigned themselves to their fates and were most unwilling to assert themselves for a change.

Ever since the time the poet took charge of his paternal acres at Shilaidaha, he devoted himself to promoting awareness among them. He wrote,

> From the verandah of the house where I lived in Shilaidaha, one sees nothing but field upon field stretching out beyond the horizon. From early dawn one peasant after another comes with his plough and cattle, ploughs round and round his tiny plots and departs. How great is the waste of divided effort! And I have seen it daily with my own eyes![140]

He was aware of the complete dependence of Indian agriculture on the vagaries of nature. He tried to explain to peasants on his estate how with a mechanical plough and harvester, they could complete cultivation and gather the ripened grain with less labour and in shorter time thereby often minimizing the damage to the reaped crop even when it still lay on the fields. The poet enumerated how their attempts to raise crops on strips of lands separated by ridges, by means of time-honoured ploughs was as good as trying to fill a 'bottomless pit'. Poverty would persist as long they continued to cultivate their lands separately and in isolation. In his understanding, most peasants in Bengal had insufficient resources that often delayed the ploughing process at great cost to the crop. Besides, much of the labour of the plough animals was wasted, as the direction of the plough changed frequently because of the twisted boundary lines of the plots. Therefore, he was convinced that

> If each cultivator did not regard his small holdings as an independent unit, if all adjoining strips were reckoned as one, fewer plough shares and bullocks could do the tilling and much wasteful labour would be eliminated. There would again be a great saving of energy and expense if after harvest the farmers collectively stored and marketed their produce.[141]

Labour-saving devices and lucrative utilization of surplus time could better their economic prospects. The poet advised cultivation of vegetables, viz., cauliflower, maize and peas of the Patna variety for domestic consumption at leisure as also adoption of practices of crop-rotation, cultivation of commercial crops, viz., sugar cane and conversion of one-crop fields into multiple cropping.[142]

However, Tagore found it very difficult to convince peasants uniformly to adopt new practices. There were those who willingly sweated over paddy but would make no effort to grow green vegetables around their homesteads while there were others in some districts that were busy all the year round with rice, jute, sugar cane, mustard and all other spring crops.[143] Response to new ideas varied from hostility to reluctant acquiescence. The poet had the experience of witnessing in some regions holdings which were not fit for such produces left fallow even while compelled to pay rent to the landlord while in the same locality came farmers from up-country who leased the fallow strips of land. They raised several varieties of melons. Bengali farmers even while aware of the annual profits of melon growers refused to cultivate it on their sandy soil. They found it hard to change the 'habits of body and mind which have grown upon him all through his years'. Therefore, the primary need was to remove the psychological obstacle. The way of knowledge was therefore the only true way. It was imperative to teach the peasant to employ all his energy in improving his own line of work.

To make scientific agriculture a reality in the Bengal countryside, Tagore emphasized the importance of easy sources of rural credit to weaken the stranglehold of moneylenders on agrarian society. Co-operative banks seemed the only solution to rampant indebtedness among peasants and artisans.[144] *Krishibanks* were to serve the twin purpose of providing loans for productive purposes at low rates and instructing ryots on curtailing unnecessary expenditure and saving. He also advised them to ensure that money borrowed for a specific purpose was utilized to that end. The poet regarded such banks as indispensable to the work of rural reconstruction.[145]

To mitigate the distress of the rural populace during natural calamities, Tagore suggested cooperative grain stores or *dharmagolas*. Each ryot was to contribute a fixed proportion of his harvested crop to help stock the *dharmagola*. The grain thus stored would help the village to combat food shortage. The ryots would also be provided with seeds at nominal rates.

Tagore was not content to restrict himself to the theoretical plane as these ideas gradually unfolded in his mind. In his lifetime, he tried his best to translate his ideas into practice in different parts of his zamindari. Not once did he endeavour to take the responsibility of the whole country. But he was convinced in his mission to hold up an ideal for his countrymen to emulate.[146] He attempted to free only a few villages from the shackles of superstitions, ignorance, backwardness and inability.[147]

Through his agricultural experiments he tried to achieve two goals: familiarize his countrymen with new techniques to initiate an increase in agricultural yield and make possible actual improvement in the material condition of the rural populace, particularly the peasantry. It was at Shilaidaha in the 1890s that Tagore first tried to apply some of his ideas.[148] Three trends can be discerned in his activities here: impart practical training in agriculture, create ideal villages and organize groups of young volunteers.[149] Experiments were made with various kinds of vegetables, fruits, silkworms, etc., as also with the soil available. Potato cultivation was undertaken. Large quantities of potato seeds were brought from Nainital at great cost. Fertilizers were used to enrich the soil. Officials of the state agricultural department helped him with seeds that were distributed among ryots on the estate.[150] Though initially his experiment failed due to faulty technique, it had considerable impact on the ryots of the neighbouring areas, many of whom were inspired to undertake potato cultivation.[151] Maize and paddy cultivation was also experimented with and high quality seeds were brought from America and Madras respectively.

On his return from America, Rathindranath set-up a farm for agricultural research on 80 *bighas* of land at Shilaidaha.[152] A research laboratory was set-up for soil testing. Workers experimented with different types of crops and fertilizers. Pump irrigation was initiated.

Agricultural circular carrying detailed information on crops and seasons was published and distributed among ryots in the neighbouring region to facilitate efficient production. Rathindranath imported machines from America for the purpose. A band of dedicated volunteers was organized to undertake cultivation of different kinds of flowers, groundnut, peas, onions, cauliflowers, etc., on the estate.

Experiments were also made with silk cultivation here. Different varieties of silkworms were tested. Rathindranath was helped in this venture by his friend Akshoy Kumar Moitra, who himself established a silk factory at Rajshahi and bestowed some twenty silkworms upon Tagore who later wrote to the scientist J.C. Bose that their number had increased to two lakhs. Ten to twelve persons were engaged in caring for them throughout the day.[153] However, this experiment with silk had to be finally abandoned, as the silk produced could not be sold. Yet, the poet did not lose hope. He encouraged his ryots to experiment with new crops depending, upon the suitability of soil conditions and convert mono-crop lands into double or multiple cropping ones. Cultivation of watermelon, *Kankur* and *Kalai* was advised on sandy soil while maize, sugar cane, cauliflower or peas could be alternated with the usual paddy. He had sugar cane of the *gandari* variety, brought from Dacca and cultivated in Shilaidaha. Rabindranath organized a group of young volunteers or *bratidal* who were imparted practical primary training in agriculture and entrusted, with the task of familiarizing the ryots in the neighbouring areas with new techniques.[154]

At Shilaidaha, it was the cooperative principle that underlay all his experiments with agriculture. A *krishibank* was organized to help peasants with soft loans with borrowed capital from the poet's friends and rich *mahajans*. A cooperative grain bank was started with each ryot's grain contribution. The latter were also supplied with seeds at low costs.

When the poet lost Shilaidaha to a member of his family after the partition of zamindari estates, his agricultural experiments were abandoned in the first decade of the twentieth century. He then shifted his focus to Patisar in Kaligram *pargana* in Rajshahi district where, work had begun earlier. But agriculture was not easy here as land was not conducive. Rains caused extensive waterlogging. So he advised the peasants to grow fruits, viz., pineapple, date palms, etc., along with potato. Rathindranath introduced tractors in 1905-6 for the first time. These attempts had significant demonstration effects. In 1905, a *krishibank* had been started here to finance agriculture and cottage industry. Borrowers were required to pay an interest of 9 per cent and use the loans for the purpose for which it had been sanctioned. Agricultural loans had to be repaid after the harvest and ryots, were most often given a remission of 3 per cent on the interest rate. Once the loan was repaid one was free to borrow again.[155] A Karmi Sangha was started to help the tenants on the estates and ryots in adjoining areas.[156] It attained some success in popularizing scientific techniques and checking moneylending operations. Irrigation works, viz., wells was undertaken. The local administration report informs:

> It must not be imagined that a powerful landlord is always oppressive and uncharitable. A striking instance to the contrary is given in the Settlement Officer's account of the estate of Rabindranath Tagore, the Bengali poet, whose fame is worldwide. It is clear that to poetical genius he adds practical and beneficial ideas of estate management, which should be an example to the local zamindars. . . . Sub-infeudation within the estate is forbidden, raiyats are not allowed to sublet on pain of ejectment. There are three subdivisions of the estate, each under a sub-manager with a staff of tahsildars, whose accounts are strictly supervised. Half of the Dakhilas are checked by an officer of the head office. Employees are expected to deal fairly with the ryots and unpopularity earns dismissal. Remissions of rent are granted when inability to pay is proved. . . . There are Lower Primary schools in each division, and at Patisar, the center of management, there is a High School with 250 students and a Charitable dispensary. These are maintained out of a fund to which the estate contributes annually Rs. 1250. . . .[157]

Tagore's attempts were thus being brought to the notice of the administration as also the Bengali literati. However, activities in Patisar soon received a setback when one of the estate workers Atul Sen and his group of volunteers aroused suspicion of the British government and were interned and the work had to be abandoned.

It was to be taken up again at another place, Sriniketan. In 1912, Rabindranath bought some lands in Surul near Santiniketan where he had started his educational institution years ago.[158] The poet expressed his dream for Sriniketan, 'The sympathy that I feel for millions of my countrymen who cannot get a square meal throughout the year even for a day will find expression in Sriniketan.'[159]

His attempts at rural development were to reach a new level here. An agricultural farm and a research laboratory were started. Agriculture and dairy work continued simultaneously. In the farm, experiments were made with cultivation of paddy, groundnut and chilies as also with fruits and vegetables, viz., lime, jackfruit, tomatoes, brinjals, etc. It was also planned to grow *napier* grass and *jowar* to combat fodder shortage during the monsoon.[160]

A new dimension was added to the efforts at Sriniketan after L.K. Elmhirst, an agricultural economist, joined in 1921. He organized a group of young volunteers who were to be trained in rural resuscitation work.[161] For Rabindranath, the two experiments, pedagogic and agricultural, cultural and rural, were vitally linked. Elmhirst's initial difficulties were great. He had to combat numerous obstacles in the shape of ignorance, lethargy, suspicion and social inhibition of various kinds all of which he overcome with love, sympathy, patience and hard work. He went to the very root of the problem as he attempted, to break down the age-old social conservatism and infuse in the rural populace a feeling of mutual respect and self-help. On 6 February 1922, a new chapter began when the Department of Rural Reconstruction and Rural Development was opened in Sriniketan. It aimed to bring under the scope of its activity the neighbouring villages. The aim was twofold: education and extension. The latter included, all activities relating to the reorganization and reconstruction of village life as well as to the building up of rural industries and organizing agricultural and business enterprises.[162] Specific programmes for the economic, social and health rehabilitation of neighbouring villages were formulated.[163] The work involved: imparting the knowledge and experience acquired in the classroom and the experimental farm to the inhabitants in the pursuit of betterment of sanitation, health and credit facilities; providing assistance in marketing their produce and procuring components of production at favourable rates; popularizing scientific methods of cultivation and animal husbandry; encouraging cottage industry and artisan production and; bringing home to them the benefits of associated life, mutual aid and common endeavour.[164]

Elmhirst was the chief architect of Sriniketan, which he built-up with his own labour and later gave financial stability to the project with funds from his own resources. The work that was started was not intended to be simply an isolated undertaking. It was a sort of movement full of possibilities that could take the shape of community development. Elmhirst stayed for three years but his agricultural experiments continued.[165] The farm that gradually grew in size over the years laid equal emphasis on experimentation, training and extension. Experiments were made with different types of crops and fertilizers and crop rotation.[166] The problems of fodder shortage, deep ploughing, manuring, trenching, conservation of moisture and introduction of mechanization were addressed.[167]

To impart training in agrarian techniques apprentices were accepted. Arrangements were made for long- and short-term programmes. Students were given projects to work on. They were taught the basics of agricultural economy, collection and analysis of relevant data, analysis and conservation of soil, selection of seeds, use of fertilizers, irrigation, protection of crops, etc. Each year, young agricultural enthusiasts and ryots from adjoining areas came to Sriniketan for training. Extension work involved acquainting ryots of nearby regions with scientific agricultural practices. 'Demonstration plots' were started in many villages. Workers from Sriniketan made regular visits to oversee the work. Ryots were provided with implements at low cost. Health and sanitation were given priority. All those who were unable to come to Sriniketan for training, could benefit from their nearest experimental firm. Regular discussions and exhibitions were held. Circulars and pamphlets giving detailed information and published from Sriniketan were circulated among the local villagers.[168]

Particular importance was given to animal husbandry. A scientific work plan was adopted for dairy, poultry and breeding of goats and bees.[169] The Sriniketan dairy provided fresh milk to Santiniketan and Sriniketan and bred good quality cattle. It was attempted to induce cultivators in the neighbourhood, to adopt scientific techniques of breeding. Poultry work begun in 1924 was highly successful. Apprentices were taken who were, later on helped to set-up their own poultry farms.[170] Breeding of bees, however, did not prove feasible and the project had to be dropped in the long run.

'Bratisangha', a group of volunteers was entrusted with the task of familiarizing the villagers with improved techniques of agriculture, cottage industry and health care. Cooperative societies covering health, irrigation, pisciculture, credit and rural development were established. In 1927, Visva-Bharati Central Cooperative Bank was opened. Agricultural loans were advanced at low rates. Attempts were made to foster cooperation among local farmers in buying, selling and for credit. A cooperative grain-store or *dharmagola* was started in 1928, to meet the grain crisis in times of famines and floods. Similar ventures were initiated in adjoining villages.

To underscore the benefits of collective agriculture, the poet initiated festivals, fairs and exhibitions.[171] In this way, he attempted to give decaying Indian villages a new lease of life by injecting new blood into the rural economy through agrarian revival. His approach was both realistic and practical. His experiments continued even after his death in 1941.[172]

TAGORE ON RURAL INDUSTRIES

Rabindranath perceived rural industrial planning as, inherently indispensable to the welfare of the rural folk and as a means for supplementing agricultural income. He was convinced that, agriculture alone could not support the ever increasing population. Besides, agricultural work was not available throughout the year. Decline of indigenous industries over the years and absence of other means of livelihood, had compounded the misery of the rural folk because of the increased pressure on land. This had reduced marginal productivity. It was thus imperative, to solve the problems of disguised and seasonal unemployment through revitalization of rural industries. From the days of the Swadeshi movement, he explained the need for developing rural handicrafts industry.[173]

His programme recognized the importance of the use of machines and application of power for increasing efficiency of production. He welcomed the development of cotton textile industry in Bombay and was at the same time, appalled by the apathy of the people of Bengal towards industrial enterprise. Bengal had once a thriving cotton textile industry, which needed to be revived in order to help the work of rural resuscitation.[174] However, he was a staunch critic of Gandhi's emphasis on hand spinning. For Gandhi *khadi* was the only sound economic proposition of India. He said:

> *Khadi* is the only true economic proposition in terms of the millions of villagers until such time, if ever, when a better system of supplying work and adequate wages for every able-bodied person above the age of sixteen, male or female, is found for his field, cottage or even factory in every one of the villages of India; or till sufficient cities are built-up to displace the villages so as to give the villages the necessary comforts and amenities that a well-regulated life demands and is entitled to. I have only to state the proposition thus fully to show that *khadi* must hold the field for any length of time that we can think of.[175]

In contrast[176] Rabindranath did not believe that revival of villages was possible by simply turning to the *charkha* and weaving homespun yarn. Spinning was not creative. He who constantly turned the wooden wheel became a machine himself. Such a person could not actualize the poet's plan of rural reconstruction. *Swaraj* for the whole country could begin only in the village where the people had organized themselves into a united community.[177] They would concentrate on improving the facilities for health, education, economic activities and amusements in villages. That *swaraj* would then advance by its own power, 'its propulsion inherent in the organic process of its own living growth', and not in the mechanical rotation of the spinning wheel. The poet was convinced that to give the *charkha* the first place in the work of national uplift was not conducive to general welfare. An image of the country's welfare in its broader aspect would draw towards it the energy of the people, fed with the best of emotion and intellect. Acts of self-sacrifice needed the stimulus of that great vision—and that 'no piles of cotton thread could possibly create'.[178] Nor did he agree that over-emphasis on *charkha* would highlight the necessity of adoption of labour-intensive production process and self-employment as a means of solving the problem of unemployment.[179]

The poet also did not undermine the importance of heavy and capital-intensive industries. He considered them indispensable for the country's economic recovery in spite of the accompanying social and moral evils, viz., concentration of wealth in the hands of a few, unequal distribution of wealth, breakdown of joint families, migration of population from the villages to urban areas as source of labour, etc.[180] But large-scale industries were costly ventures requiring elaborate infrastructure and planning. They could never bring about the kind of rural reconstruction that he envisaged. That role he assigned to cottage and handicrafts industries.

In his attempt to realize some of his ideas formulated over the years he was inspired by the attempts of his siblings and nephews at business ventures. Most of these were abortive attempts but they helped to strengthen the poet's belief in their importance in his programme for all-round rural recovery and as a supplement to agricultural revival.[181] At Shilaidaha, he tried to breed silk-worms to invigorate the indigenous silk industry. He even endeavoured to run rice mills and a pottery industry.

Sriniketan gave him a greater opportunity to experiment with rural industries. Work started in a tin shed—'Bichitra Studio'—in Santiniketan under the supervision of his daughter-in-law Pratima Devi. *Batik* printing, book-binding, lac work and such other crafts were taught. The workshop was shifted to Sriniketan in 1928 and renamed first 'Shilpa Bhavan' and then in 1951 'Shilpa Sadana'. A rural handicrafts centre (Palli Karukari Kendra) was started by Pratima Devi to impart training in stitching, knitting and similar crafts to the women in the neighbouring villages. To train artisans Shilpa Bhavan adopted a Cottage Industries Training Programme. Workers were sent everywhere to learn new techniques of production in use elsewhere in the world. This knowledge was then imparted to enable them to improve the quality of their products.[182] It was sought to establish an extension and rehabilitation wing for the artisans. It would provide them with capital and a marketing organization to help buy raw materials and sell their products.[183] A sales counter was opened in Calcutta in 1938. It was hoped that villages would not only become self-sufficient but would also cater to the demands of distant markets.[184]

Gradually, all sorts of cottage industries were initiated under the aegis of Shilpa Bhavan. In the weaving sector, *saris*, mats, blankets, etc., were woven as also silk cloth. Dyeing, calico printing and analysis of cloth were taught. After completing their training, most artisans went back and were helped to launch their own production units. Teachers were also sent to the villages to teach weavers in nearby Surul, Bahadurpur, Bhubandanga, Santalpara and Dangalpara. Villagers were taught to weave mats, wrappers, etc. Sriniketan also opened the first tannery in Birbhum district. Students came from far-off places, viz., Burma and Madras. Leather needed for shoes, suitcases, etc., were tanned here. However, it was not possible to operate the tannery for long and it had to be closed down in 1932. But production of leather goods continued.[185] Pottery work began in 1929. Machines were brought from Germany. With government aid a pottery shed and furnace were built. Pots, pans, utensils, etc., were sold in the market. Wood-work was begun under the supervision of a Japanese teacher Kashahara. Boys from nearby came to learn. Shilpa Bhavan was able to meet all the furniture requirements of Santiniketan and Sriniketan. There was a demand for toys, handlooms and other necessities produced here. *Batik* and embroidery work was also begun. Women from nearby took them up in large numbers. They were also taught cutting, stitching and knitting.[186] It was attempted to revive the old *lac* industry of Birbhum district.[187] Book-binding was started among the students of Siksha Satra, the poet's school in Sriniketan.[188] In 1936, basketry and cane-work was begun. Hand fans, mats, hats, candle-stands, etc., were made from bamboo and cane. The products of Sriniketan, popularly called 'Sriniketan Crafts' were soon in great demand.[189]

Thus professional training was offered in a number of crafts.[190] The aim of Shilpa Bhavan was primarily to equip rural artisans with the best techniques of production so that they could work independently. Cottage industries could provide a perennial source of income. Here also Tagore emphasized the need for cooperation to ensure success. It would facilitate all types of work, improve production, marketing and the bargaining capacity of workers.[191] Lady workers of Sriniketan visited the villages to train the rural womenfolk. It was hoped to equip them to work from home and thus augment the family income. The poet hoped that cottage industries would be able to provide employment, part-time and full-time, to a substantial section of the population. This would considerably lessen the pressure on agricultural sector and help in rejuvenating rural economy.

RESUSCITATION OF VILLAGES

Rabindranath's programme for rural reconstruction encompassed, remedies to the maladies afflicting rural life and included much more than economic development. He adopted a wholesome approach to issues vital to peasant life and suggested ways and means to make possible healthy living. Prescriptions for mere agrarian and rural industrial revival were not enough. As he said he wanted to attain *swaraj* in its real sense in every village—a miniature of what should be there in the whole country. He gave importance to communal harmony, education, health and self-government. Warning his countrymen about the pitfalls of communal disharmony which vitiated rural life and strained agrarian relations he suggested educating the common folk in the basic tenets of all religions.[192]

Promotion of education was central to his work. He ascribed many of the ills suffered by the peasants to their lack of basic education. He suggested establishment of schools including night schools for village children.[193] At Shilaidaha, he started a school in his estate. This experiment was to receive its complete form at Santiniketan-Sriniketan. He was also instrumental in initiating village fairs which played a vital role in the life of an average villager. They were an excellent medium for educating them and restoring Hindu-Muslim amity.[194] The 'Katyayani mela' (1905) and 'Rajrajeshwar mela' (1907) were started at Shilaidaha. At Patisar, free elementary schools were started. Primary schools were set-up in each division. There was an English high school with over 250 students.[195] Conscious efforts were made to eradicate illiteracy in the neighbouring villages. The education system envisaged at Sriniketan was intended to be work-centric from the beginning, very different from the poet's institution at Santiniketan. Rural needs were prioritized to draw the peasants of the neighbouring villages. The school Shiksha Satra, first begun in Santiniketan in 1924, was shifted to Sriniketan 2 km away in 1926.[196] General education was combined here with social welfare activities. Keeping in tune with the needs of rural life, agriculture, animal husbandry and cottage industry were encouraged. Shiksha Charcha was the training centre for rural teachers. This was shifted from Bolpur to Sriniketan in 1937. The training period was later reduced from two to one year. Lok Shiksha Samsad was started in 1937 to make possible some form of distance education. With Sriniketan as the centre, its branches were spread out in the neighbouring villages. Female and adult education was promoted. Library facilities were provided. Women were also given vocational training. Night schools were organized. To promote mass education, lantern lectures, festivals, fairs and exhibitions were arranged.[197]

Rural health and hygiene was also given top priority. Water and sanitation are essential conditions that need to be created and sustained and matters most to human well-being. Delivering clean water, removing waste water and providing sanitation are three of the basic foundations for human progress. The poet attempted to promote an awareness of health and hygiene as he experimented with rural resuscitation programmes. Hundreds of people died every year in the villages due to lack of clean water and proper sanitation. The poet considered unclean water to be a greater threat to human security than any political developments including war. He realized that lack of awareness of health and hygiene combined with the non-existence of basic medical facilities was the bane of life in most villages.[198] Disease, he said, was his greatest adversary.[199] As in all his other endeavours, even in his efforts to promote public health in the countryside, Tagore based his work on two principles: *atma-shakti* or self-reliance and the spirit of cooperation, which underlay all his work.

It was as Shilaidaha that he first experimented with promotion of rural health. He began by identifying the obstacles to healthy living. Absence of sources of pure drinking water, sanitation facilities, qualified medical practitioners and health centres to help fight some of the common diseases, viz., dysentery, cholera and malaria constituted the basic hindrances.[200] He planned the *mandali* system to encourage joint responsibility for education, health and finance. Each *mandali* was to consist of a few villages. They were to regulate all activities in the latter. This he felt would make self-government a reality. The *mandali*s were to establish health centres and promote health awareness along with other necessary work. The heads of the *mandali*s were to be responsible for planning and organization. At the poet's directive, the estate was to bear the expenses. In fact, he instituted the collection of a cess *kalyanbritti* at Shilaidaha. It was first at the rate of 3 *paisa* per rupee of the revenue. He himself contributed from the income of his zamindari an amount equal to that collected from the peasants to meet the expenses of welfare activities. This amount was spent according to the principle agreed upon by the heads of the *mandali*.[201] Amenities for safe drinking water, were generally provided through digging of wells and cleaning of existing water bodies in the villages.[202]

In keeping with the poet's conviction, the young group of volunteers was instructed to instill in the rural folk an awareness of diseases and bring the sick to the estate for medication. Faith in superstitious cures was strictly discouraged.[203] The volunteers collected information about the households, diseased members, birth and death rates in each village. Those seriously ill, were transferred to the nearest government hospitals. At Birahimpur, the poet funded the Maharshi Charitable Dispensary to cater to the immediate needs of the villagers. It had an outdoor unit.[204] Paucity of funds prevented him from opening an indoor unit though, there are instances of makeshift arrangements being made to house the critically ill.[205] Qualified doctors and compounders were appointed to provide medical help. The poet continued to fund the dispensary even after he shifted base to Kaligram in 1915.

At Kaligram, three health centres were opened at Patisar, Kamta and Ratoal. Dispensaries were attached to them and hospital arrangements made. Doctors were appointed and a few beds were provided. Medicines were distributed free of cost. Expenses were shared between the estate and the local residents, the major share being borne by the former. Here the poet was partially successful in making his *mandali* system undertake health programmes.[206] Along with imparting elementary education, undertaking welfare activities, viz., clearance of forests; digging of wells and repair of roads; extending financial help to peasants; and settling disputes through arbitration, the *mandals* also provided medical facilities. A cess *hitoishibritti* was fixed at the rate of 3 *paisa* per rupee of the revenue to meet the expenses.[207]

Rabindranath continued his work, undaunted even when he shifted his base to Birbhum district which had once been noted 'for its salubrity'.[208] But gradually it fell prey to all kinds of diseases and ailments that reigned supreme in rest of the Bengal countryside.[209] Tagore soon entered into a struggle against them. It was reported:

> When in 1922 Visva-Bharati workers started welfare work in some of the villages around Sriniketan, they soon discovered that no improvement in the condition of the villagers could be as vital as that of improving the health of the people. Since then they have been carefully studying the problem and trying out different methods of placing the benefits of medical science within reach of the poverty-stricken masses. It did not take long to realize that any scheme involving charity will defeat its own object. The problem was to devise a scheme of health work which could be maintained by the people themselves.[210]

Elmhirst organized a health programme for general medical care and the eradication of malaria.[211] He was helped by the *bratibalaks*. He identified the unavailability of safe drinking water to be one of the major hindrances. Akhil Chakraborty, an engineer trained in America, was commissioned to sink tube wells at the Institute of Rural Reconstruction.[212] A large water body 'Bhubansagar' dug by the Sinhas of Raipur, from whom the poet's father Debendranath had bought the estate, was cleaned up.

Efforts were made to fight malaria on the basis of cooperation.[213] Kalimohan Ghosh was Elmhirst's most trusted worker here. Tagore later sent him to Yugoslavia in 1930 to study rural health organization.[214] The Sriniketan dispensary was opened in 1923. Work was begun in real earnest through cooperative societies.[215] A comprehensive scheme for the introduction of cooperative health services offering both preventive and curative care was begun.[216] In fact, ten health cooperative societies were set up covering an extensive area (Table 1).

TABLE 1

Name of Society	Location (Union Board)	Date of Starting	No. of Villages under the Society	No. of Members	No. of Persons covered by Membership
Bandgora-Bolpur	Bolpur	Dec. 1932	3	132	1,012
Benuri	Ruppur	Jan. 1934	6	76	385
Bahiri	Bahiri	Aug. 1936	2	150	747
Adityapur	Taltore	Aug. 1936	6	78	432
Illambazar	Illambazar	Aug. 1936	8	74	527
Adirepara	Sainthia	Dec. 1936	15	101	515
Langulia	Khatanga	Sept. 1936	16	84	455
Jaspur	Hetampur	Oct. 1937	20	102	524
Rupaspur	Khayrasole	Oct. 1937	16	110	968
Goalpara	Taltore	1932	1	42	168
		Total	93	949	5,733

There is detailed information on the diseases that afflicted the region and the patients whose medical cure was undertaken by there health cooperatives (Table 2).[217]

The health societies thus benefited hundreds of families. The provision of medical aid, which involved great expenditure, helped to establish close contact with adjacent villages.[218]

At the initial stages, the health programme was mainly curative. The first clinic that was opened was put under the care of Miss Gretchen Green, an American woman, who was experienced in nursing and first aid. The first task was to promote health awareness. The scope of work extended with the establishment of health societies that covered many villages.[219] They were affiliated to a health union that came to be located at Sriniketan with a central hospital and a clinic attached to it. A chief medical officer supervised the work of different units and offered free consulting services to the village health societies. He was appointed and maintained by the Institute of Rural Reconstruction. Each village health society appointed its own doctor and compounder. A dispensary was attached to each. They were primarily financed from fees paid by members. Sriniketan helped these societies by providing small subsidies and ran the Calcutta Hospital and Clinic from its own budget. Medicines were provided to villagers free of cost. At the institute headquarters, the clinic received patients in its outdoor wards and also ran a cooperative insurance scheme for the members of the staff and students of Sriniketan.[220] A vital

TABLE 2

Items and Diseases	*Bolpur* Bandgora	Benuria	Bahiri	Adityapur	Illambazar	Adirepara (Started 16 Dec. 1936)	Langulia	Total
No. of visits of patients	4,332	1,180	2,001	5,339	2,896	753	627	14,548
Doctor's visit to member's house	613	237	321	236	395	76	129	2,007
Malaria	235	171	214	294	514	72	102	1,602
Kala-Ajar	3	Nil	Nil	Nil	17	1	Nil	21
Veneral diseases	26	Nil	13	46	9	1	9	109
Tuberculosis	2	Nil	Nil	5	Nil	1	Nil	8
Pneumonia	11	2	13	9	18	5	18	76
Other respiratory diseases	416	157	69	188	136	41	102	1,154
Circulatory diseases	4	Nil	7	6	5	Nil	7	29
Typhoid	3	1	15	4	6	8	Nil	37
Dysentery	47	16	16	21	27	11	6	184
Internal diseases	338	114	30	136	110	11	37	776
Cholera	1	Nil	Nil	3	Nil	6	Nil	10
Small-pox	1	Nil	Nil	1	Nil	Nil	Nil	2
Measles	12	6	26	11	12	8	Nil	75
Surgical cases	258	48	9	8	7	2	15	347
Female diseases	34	5	29	40	3	2	Nil	113
Maternity cases	25	9	1	4	9	Nil	Nil	48
Leprosy	Nil	Nil	1	Nil	Nil	Nil	Nil	1
Miscellaneous	12	25	46	133	73	54	36	379

programme of fundamental research on anti-malarial drugs was introduced at Sriniketan under Dr. Harry Timbers, an American worker. Research findings were communicated to the villagers in order to aid the total process of village recovery. This was intended to support the community health programme. Consequently, malarial affliction declined in the area and health awareness improved.

Very early in his career, the poet had determined not to leave expediently unobserved the importance of abundant sources of safe water in the villages, as water-borne diseases were frequent in most areas. At Shilaidaha, four tube wells were sunk at Jahedpur, Kandabari, Kaloa and Joarpur. The village ponds were cleaned up.[221] Efforts were made to link the village drainage system with the Padma River.[222] Similar programmes were undertaken at Kaligram. In all his endeavours the poet had to struggle against the general apathy of villagers towards such enterprises. But he was undaunted in his conviction and determined to provide drainage facilities, clean jungles and build roads. His *bratibalaks* were his enthusiastic companions. The General Welfare Society of Kaligram has to its credit substantial achievements in rural development work, viz., the dam at Talimpur, embankments at Kamta and Paroil, Debanagar Saria road and the Masikpur tank. Besides, wells were sunk in many places.[223]

At Santiniketan, a public works department was established to supervise such activities in 1912. It attained considerable success in building roads and sinking wells.[224] A competent firm of engineers was brought in and boring deep with superior drills found ample supply of water at Santiniketan.[225] The *bratibalaks* cleared jungles, filled up pits of stagnant water, kerosinized tanks to destroy larvae of mosquitoes, made drains in the villages to allow the water during the drains to run out and help the sick.[226] Later, at Sriniketan all efforts were made to promote healthy living in the neighbouring villages (Table 3).[227]

Considerable success was attained.[228]

Santiniketan-Sriniketan being located in the lateritic soil zone, attention was paid in particular to water supply. The Bhubansagar tank whose cleaning the poet undertook was spread over 85 *bighas* of land.[229] The effect of this enterprise was far-reaching. It attracted the notice of the local administration courtesy the district magistrate Benodbehari Sarkar. In October 1939, the government was enthused to enact the Bengal Tank Improvement Act (Act XV of 1939). In the following year, this Act was implemented in Birbhum, Bankura, Burdwan and Midnapur.[230]

TABLE 3

Village	Construction of roads in yds	Construction of drains in yds	Jungle Cleared in *bighas*	No. of pits filled up	Distribution of quinine in grs	Kerosene sprayed (md. sr. ch.)
Bandgora	8,643	1,702	47	26	83,072	2-7-0
Bhubandanga	800	1,300	3	12	10,930	1-10-0
Kashipur	225	120	X	X	X	0-13-8
Benuri	2,374	1,428	17	87	5,929	0-24-0
Bahadurpur	684	1,368	10	3	42,194	0-22-0
Islampur	400	800	1	4	12,385	0-20-0
Ballavpur	1,300	1,447	16	53	60,532	1-24-0
Santhalpara (4 Bastis)	460	920	3	6	X	Nil`
Goalpara	290	431	1	9	17,630	8-12-0

No study on Tagore's approach to rural health would be complete without mention of his views towards the birth control movement. While aware of the necessity of ensuring healthy lives for villagers to make possible regeneration of villages, he realized the need for controlling population growth. He expressed his opinion on issues of undesired maternity and inability to provide for one's children, 'I am of the opinion that Birth Control Movement is a great movement not only because it will save women from enforced and undesirable maternity, but because it will help the cause of peace by lessening the number of surplus population of a country scrambling for food and space outside its own rightful limits.'[231]

The poet reiterated his views years' later saying that he could comprehend the sorrow, pain and poverty associated with the birth of multiple children.[232] He attributed the economic difficulties of the country in part to increase in population growth.

Tagore's efforts to fight diseases and ensure better life for the rural populace and thus hold up a model of such work before his countrymen assumes a greater significance in view of the fact that the attention of his contemporaries was at the time focused on the nationalist struggle. Such gruelling work with prospects of no immediate success failed to capture the imagination of the nation in the period when the poet took it up as his life's mission. His repeated appeals to aspiring leaders to channellize the energies of the educated youth to constructive work were ignored. The work that he undertook preceded Gandhi's efforts in this direction by more than three decades. He was inspired by George Russell who underscored the importance of physical well-being as a precursor to constructive work and advocated the need for collective action for eradication of diseases.[233] Plagued by financial difficulties, lack of government aid and often untrained workers, Tagore persevered with his work, undaunted by failure to attain immediate success. Tagore's work did not sufficiently inspire his contemporaries for them to emulate him though he did address issues vital to the peasantry that were already in the public discourse. He regretted the fact that he had been unable to make a change in the condition of the villages. But it does not minimize the significance of his work or detract from its relevance to the contemporary world. In some respects he tried to achieve for his country a century ago what the United Nations in striving to provide for the under-developed nations *vis-à-vis* health and hygiene in the twenty-first century.[234]

Tagore's greatest beneficiary was intended to be the ryots, who constituted the bulk of the rural populace. He had hoped that the country would achieve a balanced economic growth that recognized both rural and industrial and cultural priorities. But his programme for economic revival contained no proposal for renunciation of the towns. For him attainment of *swaraj* was linked with upliftment of the decaying villages. His approach was village-centric. He was an agro-economist. He had a clear perception about the inputs and infrastructural changes, application of science and technology, economic organization of production, etc., that were required for agricultural production in a backward country. When he was referring to cooperative farming, he did not mean the Russian 'Kolkhoz', or collective farms in place of personal proprietary right over land. He suggested consolidation of small holdings and setting up of cooperative joint farms. In that case, even while retaining personal proprietorship of land it would be possible to acquire the advantages of economies of scale. Consolidation of holdings would lead to bigger farms, which would entail large-scale production. It is significant that he had developed these ideas so early. He was familiar with changes in techniques that had revolutionized agrarian production in the West and wanted Indian ryots to adopt them. Agriculture and rural industry constituted the two vital components of his rural upliftment programme. Income from the latter would sustain peasant families in agriculturally unproductive months. Rabindranath was the first intellectual in Bengal to envisage for the Bengal peasantry a comprehensive programme for economic recovery. His agrarian programme was a part of his total programme for rural welfare in which he assigned a positive role to the women folk. His proposal for vocational training for the rural women belonging to peasant families, was to enable them to be financially independent while at the same time supplement their meagre family income.

Tagore was probably the only intellectual who tried to translate his ideas on agrarian issues into practice. He realized what many of his contemporaries did not, that no real improvement in the condition of the peasants would be possible without overhauling the system of agriculture. In his own way he was a pioneer in the work that he begun. His work at Sriniketan continues even today. He was a visionary though at times some of his ideas were rather utopian as the one of benevolent zamindars encouraging village self-help efforts. For long, he had left the basic problem of land relations unaddressed. Yet he did attempt a realistic approach towards the peasant question though he did not dramatize every issue into a national crusade. Yet in spirit he was indeed a true ploughman, though he did not himself handle the plough except perhaps symbolically as in the beautiful wall-fresco by the great artist Nandalal Bose at Santiniketan.[235]

NOTES

1. Joyfully we till
   Dawn to dusk we spend in the fields
   Sun shines, rain falls, leaves quiver in bamboo groves,
   The air is filled with the fragrance of the tilled soil
   The newly sprouted green resonates through the tilled rows,
   As a new lyric of the young poet's creation
   Ecstasy runs in the sprouted paddy stalks—the entire earth looks up with a smile
   In the golden light of the autumn sun and moonlit nights.
   —Bichitra 130 (Swadesh), *Gitabitan*
2. *Aye re mora fasal kati –*
   Fasal kati, fasal kati!
   Math amader mita ore, aj tari saogate
   Moder gharer angan sara bachar bhorbe dine rate!!
   Mora nebo tari dan, tai je kati dhan,
   Tai je gahi gaan—tai je sukhe khati!!
   —Rabindranath Tagore, *Anusthanik*, 16
3. The Tagore family had extended properties that included areas in Sahajadpur *pargana* in the Pabna district, Patisar *sadar* in the Kaligram *pargana* of the Rajshahi district, Shilaidaha *sadar* in the Birahimpur *pargana* of the Nadia district as well as some tracts in the Cuttack district of Orissa. The family also owned landed estates in Kustia and Kumarkhali.

The poet's grandfather Prince Dwarkanath had greatly added to the landed possessions of the family.

4. As Maharshi Debendranath became engrossed in his spiritual pursuits, he came to depend increasingly on his sons for administering the far-flung family estates. Among the siblings only Dwijendranath performed the arduous task for a short while even though, his heart was in his literary and artistic pursuits. Satyendranath, the first Indian ICS, was always busy with his administrative career. Jyotirindranath was a poet whose main interest was music. Therefore, the onus was on the young Rabi ever since he was twenty-two years old. Ultimately, it was in 1896 that he was made legally responsible for the supervision of the estates.
   Refer to Tagore family papers document no. 69, dated 8 August 1896, Accn. no. 41, preserved in Rabindra Bhavan, Visva-Bharati, Santiniketan.
5. *Namonamo namo, sundari mamo janani bangabhumi*
   Gangar tir, snigdha samir, jiban jurale tumi
   Abarito math, gaganlalat chume tabo padodhuli—
   Chhayasunibir santir neer choto choto gramguli
   Pallabghana amrakanan, rakhaler khelageho—
   Stabdha atal dighi—kalojal nishithsheetal sneha
   Buk-bhara—madhu banger badhu jal laye jaye ghare—
   Ma balite pran kare anchan, chokhe ashe jal bhore

   —Rabindranath Tagore, *Bichitra*.
6. It was at Shilaidaha where he resided for more than a decade, mostly uninterrupted, since 1890 that he first came into direct contact with village life. His acquaintance with the rural world was rather limited prior to this. The journey that began at Shilaidaha took him first to Patisar and then Sriniketan, near Bolpur in Birbhum district. Many of his short stories, some of which rank among the best in the world that give us insightful glimpses of village life were composed at this time.
   Shilaidaha lay on the northernmost border of Nadia district being separated from Pabna by the river Padma, which formed the natural boundary between the two districts.
7. Rabindranath Tagore, 'City and Village', *Towards Universal Man*, Visva-Bharati, Calcutta, 1961, pp. 317-18.
8. Rathindranath Tagore, *On the Edges of Time*, Visva-Bharati, Calcutta, 1981, p. 34.
9. Bipasha Raha, 'Economic Thought of Rabindranath Tagore', *The Calcutta Historical Journal*, vol. 25, no. 1, January-June 2005, pp. 2-3.
   N.B. Songs, short stories, travel diaries, poems, novels, essays and letters written in a literary career that spanned more than half a century help us to understand the poet's perception of the rural world in colonial Bengal.

10. Rabindranath Tagore, 'City and Village', op. cit.
11. Bipasha Raha, 'Rabindranath and Rural Reconstruction Work at Kaligram', *The Quarterly Review of Historical Studies*, vol. XLIV, April-September 2004, nos. 1 & 2, pp. 104-5.
    Rabindranath described in the essays compiled in *Palliprakriti*, innumerable letters, *Kalantar*, *Religion of Man* and the short stories his varied experiences in the years he spent in the countryside.
12. Rabindranath Tagore, 'City and Village', op. cit.
13. Letter written from Shilaidaha to Amiya Chakraborty on 10 March 1893, *Rabindra Rachanbali* (henceforth *RR*), centenary edn., vol. II, p. 107. Ref. *Chinnapatrabali* III (dated 21 August 1893) and 116 (dated 21 March 1894).
14. 'Ramkanaier Nirbuddhita', 'Khokababur Pratyabartan' and 'Didi' are some of his short stories included in *Galpaguchcha* that give us glimpses into the day-to-day lives of rustic dwellers and the sad plight of the common folk. They were exploited on the one hand by the government officials and on the other, by the local power groups', viz., zamindars, *kutcherry* officials and the moneylenders. In 'Megh o Roudro', the poet depicts how respectable villagers were ill-treated by their English masters. Harakumar, the protagonist of the story, a one-time *pattanidar* who later became a *naib* of an English zamindar was once insulted by a low caste sweeper at the orders of the latter for a minor offence. This had grave social repercussions. However, the most lucid description of the unimaginable poverty, helplessness, internal rivalries and unending quarrels, which were all a part of village life, is to be found in 'Shasti'. Refer: Rabindranath Tagore, *Galpaguchcha*, vols. I-IV, Visva-Bharati, Calcutta, 1987.
15. 18.99 per cent of the Congress delegates in sessions between 1902 and 1909 were landlords.
16. Sekhar Bandyopadhyay, *From Plassey to Partition: A History of Modern India*, Orient Longman, New Delhi, 2004, p. 232.
17. S.R. Mehrotra, *The Emergence of the Indian National Congress*, Vikas, Delhi, 1971, p. 412.
18. Sekhar Bandyopadhyay, op. cit., p. 233.
19. The work for which Dutt is best known today was accomplished after his premature retirement from government service in 1897, at the age of 49, and during a period of intense activity as a writer of economic history and as a politician, who tried, by methods of persuasion and propaganda, to alleviate the misery of poor Indian peasants. During his official career in civil service, with the exception of his earlier work *Peasantry of Bengal*, Dutt devoted himself to the enrichment of Bengali literature by writing historical novels. It was only after retirement that his scholarly works *Open Letters to Lord Curzon: On Famines and Land*

Assessments in India (1900) and *Economic History of India* (vol. 1, 1900 and vol. 2, 1902) were written.

20. What Dutt said in his *Economic History of India* was presaged by a book titled *England and India* published in 1897.
21. The devastating famines of 1877-8, 1892, 1897 and 1900 occurred in the 'era of imperialism'. R.C. Dutt, *Economic History of India*, vol. 2, *In the Victorian Age*, Indian rpt. 1976, Chapter II.
22. Dutt suggested that the Englishman should ask himself the following questions:
'Does agriculture flourish? Are industries and manufactures in a prosperous condition? Are the finances properly administered, so as to bring back to the people an adequate return for the taxes paid by them? Are the sources of national wealth widened by a government anxious for the material welfare of the people?' Ibid.
23. Ibid. Refer to B.N. Ganguli, *Indian Economic Thought: Nineteenth Century Perspectives*, MacGraw Hill, Delhi, 1977, pp. 184-8 for a detailed study on the economic thought of R.C. Dutt.
24. Dutt, op. cit., Preface.
25. According to Dutt it was neither over population nor natural improvidence of the cultivator nor the extortion of the moneylenders that was the real cause of the poverty of the agricultural population. Population did not increase faster than the area of cultivation. It was not the natural improvidence of the cultivator for those who knew the Indian cultivator would agree that with all his ignorance and superstition the latter was as provident, as frugal and as shrewd in matters of his own interest as the cultivator in any other part of the globe. Nor did Dutt believe that it was the fraudulent practices of the indigenous moneylenders that were primarily responsible for chronic state of indebtedness of the actual tillers of the soil. Ref. R.C. Dutt, 'Presidential Speech', 27 December 1899.
26. Ganguli, op. cit., p. 184.
27. Saharanpur Rules (1855).
28. Dutt wrote four letters to Viceroy Curzon on the matter of famines in the country which were subsequently published in 1900 as *Open Letters to Lord Curzan on Famines and Land Assessments in India*. Refer to R.C. Dutt, Preface, *Economic History of India*, vol. 2.
29. R.C. Dutt, Letter to *The Pioneer*, 12 March 1902.
30. Ibid.
31. Ibid.
32. Ibid., 28 March 1902.
33. Act X of 1859, Bengal.
The Clauses of the Act have been discussed at length earlier.
34. Ganguli, op. cit., pp. 94-5.

35. Ibid.
36. R.C. Dutt, 'India Agriculture', in *Speeches and Papers on Indian Questions 1891 and 1901*, pp. 98-100.
37. Dutt observed:

    'The material well-being of the people . . . depends on successful agriculture, on flourishing industries, and on sound system of finance . . . four-fifths of the population of India depend directly or indirectly on agriculture. It is the main industry of India, the main source of subsistence for the people. . . . If agriculture flourishes, if the crops are safeguarded, if the land is moderately taxed, the people are prosperous. If any of these conditions is wanting, the people must necessarily be on the verge of starvation and must perish in years of bad harvest.'

    R.C. Dutt, 'The Economic Condition of India' (Speech delivered at the Philosophical Institution, Glasgow on 4 September 1901) in *Speeches and Papers on Indian Questions, 1901 and 1902*, p. 70.
38. Dutt, *Famines in India*, op. cit., 1900.
39. Dutt, *The Economic History of India under Early British Rule*, op. cit.
40. R.C. Dutt, 'The Indian Land Question', *Indian Review*, October 1902, p. 4.
41. Ganguli, op. cit.; Bhabatosh Dutt, *Arthanitir Pathe*, Calcutta, 1977, pp. 56-7.
42. R.C. Dutt, 'Presidential Speech', 27 December 1899, Lucknow Congress, in his *Speeches and Papers on Indian Questions, 1897-1900*, p. 125.
43. Dutt, 'The Indian Land Question', op. cit., pp. 4-5.
44. Ganguli, op. cit.; Bhabatosh Dutt, op. cit.
45. In Bakharganj district for instance, where Dutt was posted as collector from 1883-5, there were about 50-2 intermediaries between the zamindar and the actual tiller. Though the zamindar prospered because of extension of cultivation in hitherto uncultivated land, the cultivators were impoverished.
46. Amalendu De, *Chirasthayi Bamdobasto O Bangali Buddhijibi*, Ratna Prakashan, Calcutta, 1981, p. 20.
47. Dutt, 'Presidential Speech', op. cit.
48. Dutt, *Economic History of India*, op. cit., vol. 1, Chap. V.
49. Bhabatosh Dutt, op. cit.
50. Published in 1890. Meer Massaraf Hossain, 'Udasin Pathiker Maner Katha', in Bishnu Basu (ed.), *Meer Masarraf Hossain Rachanasamgraha*, op. cit., vol. I, pp. 120-301.
51. Ibid., pp. 128-9. The tale of Kinney's misdeeds was widely publicized by Harinath Majumdar in the columns of *Grambarta Prakashika*.
52. Ibid., p. 264.
53. Ibid., p. 133.

54. It was published in 1891. Refer to Sufia Ahmed, *Muslim Community in Bengal, 1884-1912*, Oxford University Press, Dacca, 1974, p. 348.
55. N.A.K. Yusufzahi, *Bangiya Mussalman*, p. 16, cited in ibid.
56. Muhammad Mashenullah, 'Budir Suta', pt. II, pp. 35-6; Sufia Ahmed, op. cit.
57. Rafiuddin Ahmed, *The Bengal Muslims, 1847-1906: A Quest for Identity*, Oxford University Press, Delhi, 1987, p. 139.
58. General Report on Public Instruction in the Lower Provinces of Bengal: Report of the Director of Public Instruction 1871-72, p. 25.
59. Kazi Abdul Mannan, *Adhunik Bangla Sahitye Muslim Sadhana*, University of Rajshahi Press, Rajshahi, 1961, vol. I, pp. 123-41.
60. Sheikh Abdus Sobhan, *Hindu Mussalman*, Calcutta, 1888, pp. 2-4, cited in ibid.
61. Ibid., p. 131.
62. Ibid.
63. Muhammad Mashenullah, op. cit.
    Ref. Sufia Ahmed, op. cit.
    Mashenullah created a ryot Guru Pramanik, who conveying the sentiments of the author, blamed the government equally for the miserable condition of the peasants.
64. It was established by Abdul Lateef in 1863.
65. It was established by Syed Ameer Ali in 1877.
66. They were opened in 1883, 1890 and 1891 respectively.
67. Mirza (1840-1913) was the first Bengali Muslim to graduate from the Calcutta University. He joined Ameer Ali's Central National Muhammedan Association of Calcutta.
68. Delawarr Hussain Ahmed Mirza, *Essays on Muhammedan Social Reform*, vols. I & II, Calcutta, 1889.
69. Bipasha Raha, 'Economic Thought of Rabindranath Tagore', op. cit., pp. 2-3.
70. Rabindranath Tagore, 'City and Village', op. cit.
71. Ibid.
72. Rabindranath Tagore, *Galpaguchcha*, vols. I-IV, Visva-Bharati, Calcutta, 1987.
73. Rabindranath Tagore, *Chinnapatrabali*, III (21 August 1893) and 116 (21 March 1894).
74. In the period from 1892-3 to 1902-3 he lived mainly in the countryside. The majority of the stories included in *Galpaguchcha* were written at that time.
75. Letter to Amiyachandra Chakraborty. 15 November 1934, *Palliprakriti*, Visva-Bharati, Calcutta, 1987, p. 196.
76. Ibid., p. 6.
77. Ibid., pp. 95-7.

78. Alok Banerjee, 'Tagore on Rural Community Development in India', in Santoshchandra Sengupta (ed.), *Rabindranath Tagore: Homage from Visva-Bharati*, Visva-Bharati, Santiniketan, 1962, pp. 215-17.
79. Tarashankar Bandyopadhyay, *Rabindranath O Banglar Palli*, Mitra & Ghosh, Calcutta, 1971, p. 38.
80. Rabindranath Tagore, 'Swadeshi Samaj', 'Atma-Shakti', *RR*, vol. III, op. cit., pp. 513-635
81. Tapati Dasgupta, *Social Thought of Rabindranath Tagore: A Historical Analysis*, Abhinav Publication, New Delhi, 1993, pp. 115-16.
82. Rabindranath Tagore, *The Co-operative Principle*, compiled by Pulinbehari Sen, Visva-Bharati, Calcutta, 1963, p. 9.
83. Rabindranath Tagore, 'Swaraj Sadhan', *Kalantar*, Visva-Bharati, Calcutta, 1344 B.S.
84. Rabindranath Tagore, 'City and Village', op. cit., p. 320.
85. Ibid. The poet narrates many other similar cases. He had built a road from his estate office to Kustia. He told villagers who lived close to the road that its upkeep was their responsibility. They could easily combine and repair the ruts. In fact, it was their ox-cart wheels that damaged the road and made it unstable during the rains. But they said, 'Must we look after the road so that gentle folk from Kustia can come and go with ease?' They could not, observed the poet, bear the thought that others should also enjoy the fruits of their labour. They would much rather suffer all inconveniences.
86. Rabindranath Tagore, 'Presidential Address', *Towards Universal Man*, op. cit., pp. 123-4.
87. Rabindranath Tagore, *Palliprakriti*, op. cit., pp. 555-9.
88. Ibid., pp. 222-4.
89. The poet visited Russia in 1930. This visit had a phenomenal impact on his understanding of rural needs.
90. *RR*, op. cit., vol. 12, p. 813.
91. He wrote in an essay 'Mookhujjey Banam Banujjey', immediately after taking charge of his zamindari, that it was the zamindars who alone would provide leadership, ibid.
92. Rabindranath Tagore, 'Presidential Address', *Towards Universal Man*, op. cit., pp. 123-4.
93. Ibid.
94. Prabhat Kumar Mukhopadhyay, *Rabindrajibani*, Visva-Bharati, Calcutta, 1961, 4 vols., published between 1340 and 1363 B.S.
95. The poet expressed his view in clear terms in the rejoinder that he wrote to Pramatha Chaudhuri's essay, 'Rayater Katha'. *Sabujpatra*, Asar 1333 B.S.
96. Satyendranath was the first Indian ICS. He was for long involved with government service and posted in different places.

97. Dwijendranath was a philosopher and poet, Jyotirindranath a talented musician and Swarnakumari Devi was the first woman novelist. Apart from her creative writing she was also editor of *Bharati*. The poet's nephews Abanindranath and Gaganendranath were highly acclaimed artists with their own distinctive styles and made phenomenal contributions to the Bengal school of art.
98. Though silent on the latter issues, Tagore clearly expressed his stand on the former in his rejoinder.
N.B. Pramatha Chaudhuri, op. cit.
99. Pramatha Chaudhuri (1868-1946) was the husband of the poet's favourite niece Indira. He was a prolific writer whose essays and short stories were highly acclaimed. Born in an aristocratic family, he imbibed the culture of English educated metropolitan elite and went abroad for higher education. On his return, he devoted himself to literary pursuits. The bulk of his works were produced after 1913. He edited the popular journal *Sabujpatra*. He introduced a new style in Bengali prose. His writings bear the stamp of a progressive mind. He was a critic of decadent society and social beliefs. His essays and short stories are partly a statement of socio-economic beliefs. Pramatha Chaudhuri, *Prabandha Samgraha,* Visva-Bharati, Calcutta, vol. I, 1952, vol. II, 1954.
100. Pramatha Chaudhuri, '*Rayater Katha*', Calcutta, Visva-Bharati. 1947. Both Rabindranath's note and Pramatha's rejoinders were published in *Sabujpatra,* Asar 1333 B.S.
101. Pramatha Chaudhuri, 'Tika', *Sabujpatra,* Asar 1333 B.S.
102. Pramatha Chaudhuri, 'Rayater Katha', op. cit.
103. Ibid.
104. Bipasha Raha, 'Rabindranath and Rural Reconstruction Work at Kaligram', op. cit.
105. Refer Raja Rammohun Roy's evidence before the British Parliamentary Commission, 1831.
106. 'Rayater Katha', *RR*, vol. 13, p. 372.
107. Achinta Kumar Sengupta, *Kallol Yuga,* DM Library, Calcutta, 1372 B.S., p. 185.
108. Ibid.
109. Rabindranath made this observation in 1928. Rabindranath Tagore, *The Cooperative Principle*, op. cit., p. 7.
110. Ibid.
111. Tagore wrote, 'Sandy soil yields no crop because it does not stick together. Sap cannot collect in that soil but escapes through the interstices. The soil can be made productive by the addition of loam, humus and leaf-mould which close the gaps. It is same with human beings where the chasms between them are wide, there strength is of no use'. Ibid., pp. 9-10.

112. Ibid.
113. The poet argued, 'This combination of many people to earn a living is known in Europe as the cooperative system. It is by this system that our country can be rescued from its age-long poverty and stagnation.' Ibid.
114. The poet cited innumerable instances of the advantages of the co-operative system.
'No one by his unaided efforts could ever hope to send a letter from Calcutta to Cape Comorin at a cost of only a few pice. The postal system is the result of a gigantic cooperative effort; its benefits are so great that in the matter of dispatching and receiving letters it gives a poor person the same facilities as a millionaire.' Ibid., p. 15.
115. That was why, the poet felt, it was man's own deficiency, and no decree of fate, that made him sink into the depths. To think there was no escape from preordained misery was to make the misery perpetual. To seek new paths in a constant renewal of strength—that had always been the secret of progress when a man waited helplessly for a turn in the wheel of fortune, he had to be regarded as shorn of manhood.
116. Bipasha Raha, 'Rabindranath Tagore and the Co-operative Idea', *Bengal Past and Present*, vol. 119, 2000, p. 112.
117. He lost a large sum of his Nobel Prize money when his cooperative bank collapsed.
118. Prabhat Kumar Mukhopadhyay, op. cit., vol. 2, p. 146.
119. In the Preface to Tagore's *The Cooperative Principle*, Sudhir Ranjan Das comments:
'It is not necessary to explain to enlightened readers the virtues and advantage of cooperative enterprise. Its usefulness is now taken for granted and serious students of rural economies are engaged in very useful research work in this field of national welfare. But over half a century ago when nobody bothered about the principles of cooperation or of their application to the rural problems Gurudeva Rabindranath Tagore thought about them and devotedly worked in this field of study as a pioneer for the uplift of countless men and women residing in remote villages scattered all over Bengal and wallowing in the mire of poverty, ignorance and superstition.' Rabindranath Tagore, *The Cooperative Principle*, op. cit., Preface, p. 5.
120. Subrata Gupta and Parimal Chakraborty, *Rabindranather Arthanaitik Chinta*, Modern Column, Calcutta, 1987, p. 102.
121. Dated 18 March 1906.
122. Rabindranath Tagore, *The Cooperative Principle*, op. cit., p. 61. Between 1918 to 1925 the poet put to paper much of his thoughts on co-operation in elaborate essays. These were later collected in a book

Samavayaniti. Besides from 1918, the Bengal Cooperative Organization Society began to publish a journal, *Bhandar,* to popularize the idea among villagers. In its first, issue (July-August 1918) Tagore's essay 'Samavaya' was published.

123. Ibid., p. 55.
124. 'It was while some of us were thinking of the ways and means of adopting this principle in our institution that I came across the book called the *National Being* written by that Irish idealist, A.E., who has a rare combination in himself of poetry and practical wisdom. There I could see a great concrete realization of the cooperative living of my dreams.' Ibid., p. 50.
125. Tagore met Plunkett on 22 July 1920, in the course of his tour of England, at the latter's flat in Mayfair. They discussed the cooperative movement in Ireland and the possibilities of its success in India. Plunkett held that it was in the economic sphere that men could really unite and cooperate. He advised the poet on the necessity of surveying local conditions to realize the actual needs of the people and then to undertake schemes gradually and warned him about failures that could have a demoralizing effect. Rathindranath Tagore, *On the Edges of Time*, Visva-Bharati, Calcutta, 1981, pp. 117-18.
126. Rabindranath Tagore, *The Cooperative Principle*, op. cit., p. 26. The poet said this in his presidential address at a public meeting held to celebrate the International Cooperators Day, 2 July 1927, under the Bengal Cooperative Organization Society in Calcutta.
127. Ibid., p. 49.
128. Revenue, Agriculture, nos. 13-14, January 1884.
129. Deepak Kumar, *Science and the Raj: A Study of British India*, 2nd edn., Oxford University Press, Delhi, 2006, p. 96.
130. J.A. Voelcker, *Report on the Improvement of Indian Agriculture* (2nd edn.), Calcutta, 1897, cited in ibid., p. 97.
131. Nityagopal Mukherjee, *Handbook of Sericulture*, Pub. n.m., Calcutta, 1899.
132. Ibid.
133. Ibid.
134. Rabindranath Tagore, *Palliprakriti*, op. cit.
135. Rabindranath Tagore, 'Bhumilakshmi', Aswin 1325 B.S., *Palliprakriti*, op. cit., pp. 37-9.
136. The poet's son Rathindranath and his friend's son Santoshchandra Majumdar went in 1906 and were joined a year later by Nagendranath Gangopadhyay who was the youngest son-in-law of Rabindranath.
137. Rathindranath informs us:

    'He (Rabindranath) thought that in order to resuscitate rural life, agriculture, which is the basic economic resource of the people must be improved. He therefore, desired that Santosh and I must go abroad to get technical training in agriculture and animal husbandry so that after our return we could help him.' Rathindranath Tagore, *On the Edges of Time*, Visva-Bharati, Calcutta, 1958, p. 135.

138. Rabindranath Tagore, 'Presidential Address', Pabna Regional Conference, Falgun 1314 B.S., *Palliprakriti*, op. cit., p. 2.

139. The poet elaborated:
'Many labour-saving appliances have been invented in Europe but the smallness of our holdings and our lack of resources makes them almost useless to us. If the farmers in a village or, better still in a community unit combined and engaged in joint cultivation of their land on a cooperative basis, they could all profit by the use of these modern machines, which would reduce expenditure and give larger yields. It is economical for them to buy even an expensive machine if all the sugar cane in the village is crushed by it. They can afford to have a jute press in the village if the produce of all the fields and homesteads is brought to a common centre. There would be an improvement in animal husbandry if all the milkmen in the village combined. Similarly, weavers in the village can indent for improved power looms if they pool their resources and work on a cooperative basis'. Rabindranath Tagore, 'Presidential Address', *Towards Universal Man*, op. cit., p. 119.

140. Rabindranath Tagore, *Letters from Russia*, Visva-Bharati, Calcutta, 1984, pp. 21-2.

141. Rabindranath Tagore, 'Cooperation', *Towards Universal Man*, op. cit., pp. 325-6.

142. Sachindranath Adhikari, *Shilaidaha O Rabindranth*, Jijnasa, Calcutta, 194, pp. 103-5.

143. Rabindranath Tagore, 'The Striving for Swaraj', *Towards Universal Man*, op. cit., p. 277.

144. Tagore wrote that peasants borrowed to start the process of production, pay rent and meet daily expenses throughout the year particularly when agricultural work was not available. Rural artisans also depended on them for the same reasons. Moneylenders charged very high interests and land once mortgaged was lost to peasants.

145. Letter written to friend Prasanta Chandra Mahalanobis, the great statistician, *Desh*, Asar 1382 B.S., letter number 60.

146. Letter to Ajit Kumar Chakraborty, 14 January 1908. Prabhat Kumar Mukhopadhyay, op. cit., vol. 4, p. 318.

147. Rabindranath Tagore, 'Sriniketaner Itihas O Adarsha', *Palliprakriti*, op. cit., p. 104.

148. He was enthusiastically supported by his friends Dwijendralal Roy, the popular poet, who was also an agricultural expert and Acharya Jagadish Chandra Bose, the eminent scientist. They were frequent visitors at Shilaidaha. Prolonged discussions were held on new techniques.

149. Sachindranath Adhikari, op. cit., p. 71.
150. It was mentioned in official records:
'Experiments with Nainital potatoes were made by Mr. Rabindranath Tagore in the Tagore estate of Shilaidaha in Kustia sub-division. The crop was not satisfactory owing to the defective cultivation. One of Mr. Tagore's tenants however working under more favourable circumstances obtained a bumper crop from a portion of the same seed and . . . the experiment is said to have induced several neighbouring ryots to take the potato cultivation.' *Land Records of Agriculture*, 1899.
151. Rabindranath Tagore, 'Asramer Rup O Bikash', *RR*, op. cit., vol. II, p. 735. Sachindranath Adhikari, op. cit., pp. 424-5.
152. Rathindranath returned from Illinois in 1909 with a Bachelor of Science degree in agriculture and animal husbandry. Tagore was then in Santiniketan. Describing his work, Rathindranath writes:
'I settled down at Shelaidah and led the life of a country gentleman. A farm was laid out, seeds of maize, clover and alfalfa were imported from America; discs, harrows and such modern implement suitable to Indian conditions were introduced. Even a small laboratory was fitted up for soil testing. The highest compliment was given me by Myron Phelps when he told me that at Shelaidah he had discovered a genuinely successful American farm.' Rathindranath Tagore, *On the Edges of Time*, op. cit., p. 86.
153. Letter written to Jagadish Chandra Bose from Shilaidaha, 10 Asar, 1306 B.S., *Chithipatra*, op. cit., vol. 6, pp. 3-4.
154. Sachindranath Adhikari, who was then a student, writes how different kinds of fruits and flowers were grown scientifically. Each student was in charge of five trees in the land adjoining the ***kuthibari***. They had to water them and apply fertilizers. In another plot adjoining the ***kuthibari*** they grew groundnut, potato, onion, peas and cauliflowers. Sachindranath Adhikari, op. cit., p. 68.
155. The poet contributed a considerable sum from his Nobel Prize money to build-up the capital of the bank; though he later lost a huge sum when the bank collapsed he was undaunted. His faith in the need and indispensability of such enterprises was unshaken.
156. Atul Sen was in charge of it. The ***krishibank*** sanctioned loans only on the recommendations of the Karmi Sangha. After harvest, the crop was brought straight from the fields to the estate office. It was only after adjusting the debt payment that the peasant could take away his share. Sudhir Sen, *Rabindranath Tagore on Rural Reconstruction*, Visva-Bharati, Calcutta, 1978, pp. 97-8.
157. L.S.S. O'Malley, *Bengal District Gazetteers, Rajshahi*, 1916.
158. If Santiniketan was the ideal of education and culture, Sriniketan was that of rural reconstruction. Both combined to form Visva-Bharati.

Rathindranath and Nagendranath were sent to initiate work at Sriniketan where they were joined later by Santosh Majumdar in 1917.

Sriniketan originally formed a part of Surul which was a commercial centre under the English East India Company. The *kuthibari* in Surul was built by Mr. Wilson. Later this *kuthibari* and the adjoining lands were bought by the Sinhas of Raipur. In October 1912, Rabindranath bought these lands, which was very near Santiniketan, from Colonel Narendra Prasad Sinha in London.

159. Sisir Mukhopadhyay, 'Rabindranather Sriniketan: Pragatishil Krishibhabnar Adipeeth', *Udichi*, Pous 1388 B.S., pp. 59-60.

160. Cultivation of fruits and vegetables was supervised by Mr. Kashahara, a Japanese teacher who taught agriculture and woodwork.

161. His contribution is an integral part of the history of Sriniketan. One can hardly think of Sriniketan without recalling the debt it owes to his dynamic zeal, devotion and initiative. He had met the poet abroad and was so inspired by the latter's mission that he wanted to be a part of it. He did the spade work, literally too, and helped the poet lay the foundation of a complex of activities, seemingly rustic and lowly, but vital for national rejuvenation in the long run.

162. P.C. Lal, *Reconstruction and Education in Rural India*, London, 1932, p. 233.

163. Satyadas Chakraborty, *Sriniketaner Gorar Katha*, Sahitya Samaj, Sriniketan, 1985, p. 14

164. *Visva-Bharati Bulletin*, no. 6, July 1928.

165. Kalimohan Ghosh and Dhirananda Roy did yeomen service for the success of the project.

166. Different kinds of paddy were experimented with, viz., *basmanik*, *paramannashal*, *raghushal*, *nonarmashal*, *sindurmukhi*, *chapshal*, *badkalamkati-65*, *jingheshal*, *basmati*. *Visva-Bharati Bulletin*, no. 11, December 1946.

167. To encourage sugar cane cultivation and to demonstrate how it could substantially increase agricultural income, the *hadi* system of making *gur* (jaggery) was introduced. Cultivation of tomatoes, soyabean, cauliflower, beetroot, beans and potatoes were tested as also motihari tobacco, Agartala *kalai* and Co213Co421 sugar cane. Fruits, viz., papaya, banana, guava and pineapple were grown. To provide fodder, *kalai*, napier grass and *jowar* cultivation was promoted. 'Silo' method of fodder preservation was adopted. In 1931, cultivation of camphor, cardamom, clove and cinnamon was introduced. Organic fertilizers were used to increase soil fertility. Chakraborty, Satyadas, op. cit., p. 65.

168. Rabindranath Tagore, *Palliprakriti*, op. cit., pp. 117-18.

169. Tagore always felt that agriculture was dependent on good cattle.

Nagendranath was told to visit Ireland to study the techniques of dairy farming practised there. Ibid., pp. 233-4.

170. Kalimohan Ghosh, 'Sriniketaner Pallisangathan', *Bhandar*, May 1975 (rpt.), p. 18.
171. 6 February being the foundation day of Sriniketan is celebrated as 'Magh Mela' through exhibitions, meals and cultural events held over three days. In 1928, Tagore introduced ceremonies, viz., 'Briksharopana' or tree planting and 'Halakarshan' or tilling with the plough. He himself inaugurated it by ceremoniously wielding the plough. Now his death anniversary is remembered through this festival. 'Pous Utsav' or winter festival was a supplementary of 'Halakarshan'. In the latter, the seeds were sown, while in the former the crop was harvested.
172. Subrata Gupta, and Parimal Chakraborty, op. cit., Calcutta, 1987, pp. 68-9.
173. In the highly acclaimed novel *Ghare Baire*, the poet describes how the protagonist of the novel, Nikhilesh, tried earnestly to rejuvenate both agriculture and cottage industries. He even undertook to manufacture sugar from the juice of date palm trees. He rendered financial help to all those who undertook similar ventures even if he realized that all were not fated to succeed. Ignoring better quality foreign products, a wealthy zamindar Nikhilesh even used indigenous goods in his day-to-day life, viz., pencils, pens, pencil-cutters, brass water pots and even candles to study at night. Rabindranath Tagore, *RR*, vol. 4, p. 481. In 'Pabna Pradeshik Sammelanir Sabhapatir Abhibhasan', 'Bangalir Kapader Karkhana O Hater Tant' and other essays Tagore expressed similar views on the need to develop rural industries.
174. Rabindranath Tagore, 'Bangalir Kapader Karkhana O Hater Tant', *RR*, vol. 13, pp. 562-5.
175. M.K. Gandhi, 'Is khadi Economically Sound?', *The Collected Works of Mahatma Gandhi*, New Delhi, Publications Division (1958-), vol. 63, pp. 77-8.
176. Partha Chatterjee, 'Nationalist thought and the Colonial World: A Derivative Discourse', op. cit., pp. 117-18.
177. Rabindranath Tagore, 'The Striving for Swaraj', op. cit., p. 284. The essay was first published in 1925.
178. Ibid., pp. 28-81.
179. Rabindranath Tagore, 'Charka', *Kalantar*, Visva-Bharati, Calcutta, 1989, p. 260.
180. Tagore welcomed the generous endowment made by Sir Jamshedji Tata for setting up modern scientific research laboratories so that the educated youth could acquire such knowledge. He even appealed to the Indian National Congress to play an active role in undertaking similar ventures in the country. Only then could Tata's dream of

scientific and industrial development be realized. Rabindranath Tagore, 'Atma-Shakti O Samuha', *RR*, op. cit., vol. 12, p. 867.

181. Rabindranath's brother Jyotirindranath started a steamer service which tried unsuccessfully to compete with the government aided 'Flotilla Company'. It had to be closed down after one of his ships sunk. The poet joined the business enterprise Tagore Company, started by Surendranath and Balendranath. During the Swadeshi Movement he started a weaving school in Kustia with Surendranath and Gaganendranath. All these failed at great cost. Prabhat Kumar Mukhopadhyay, op. cit., vol. 1, p. 358.
182. Rathindranath and Pratima Devi went to Italy to learn leather work and Manindranath Sen to Japan for weaving. The techniques of Manipuri weavers and artists of Java were also adopted. Nandalal Bose and his students in Kala Bhavan constantly provided artisans with new designs to work upon. In 1936 the number of trainees in Shilpa Bhavan numbered 109. Sugata Das Gupta, *A Poet and a Plan: Tagore's Experiments in Rural Reconstruction,* Thacker Spink, Calcutta, 1933, pp. 49-50.
183. Ibid.
184. Satyadas Chakraborty, op. cit., p. 44.
185. Products included shoes, suitcases, leather stools, wallets, purses, handbags, belts, etc. By 1940, the trainees numbered 63, many of whom were women. Gradually all set up their own production units.
186. In 1933-4, 47 local women learnt such work here.
187. Pratima Devi brought some artists from neighbouring Illambazar, a region once famous for its lac products, to Sriniketan to teach the craft here. Products included lac work on wooden goods, toys and dolls made from lac.
188. In 1930, there were ten trainees. An expert was brought in 1930 from Calcutta. All the books of Visva-Bharati were bound here.
189. Satyadas Chakraborty, op. cit., p. 36.
190. In 1930, a small power house and some machines were installed for metal and wood work. In the machine shop cart wheels, iron ovens and other household goods were manufactured. Students were trained in smithy work. They were taught to operate tractors and oil engines and undertake repairing work. In 1941, fruit preservation and canning were introduced.
191. In Sriniketan, the cooperative bank provided artisans with easy credit. Attempts were made to organize cooperative societies and run industrial centres in the villages with rural artisans. Such centres were started in neighbouring villages of Laldaha, Illambazar, Adityapur, Goalpara, Bandgora, Surul and Koridha.
192. In 1908, in a meeting of Bangiya Pradeshik Sanmelani at Pabna he

outlined a 15-point programme to be undertaken in each district. Rabindranath Tagore, *Palliprakriti*, op. cit., pp. 222-4.

193. Rabindranath's speeches in the Pabna Pradeshik Sanmelani in 1908 and Bangiya Hitosadhan Mandali in 1915 gives an insight into the evolution of his thought. He delivered these speeches on 26 January, 13 February and 28 March 1915 to mark the inauguration of the latter established in 1915 by Dwijendralal Moitra. The last two speeches were included in *Palliprakriti* as 'Karmayajna' and 'Pallir Unnati'. In all three, he elaborately discussed a programme of rural welfare activities:
    (a) Provide elementary education and basic mathematical knowledge to the illiterate.
    (b) Hold classes and distribute works on health care and nursing.
    (c) Take preventive measures against tuberculosis and malaria.
    (d) Find ways to check infant mortality.
    (e) Provide for clean drinking water in the village.
    (f) Set-up loan advancing societies in the village and make villagers aware of their benefits.
    (g) Help the poor in times of famine, flood and epidemics.
194. Rabindranath Tagore, 'Swadeshi Samaj', op. cit., p. 68.
195. Extract from L.S.S. O'Malley, *Bengal District Gazetteer* (1916) quoted in Sachindranath Adhikari, op. cit., p. 221.
196. Based on the ideas of Elmhirst, various schemes of rural reconstruction were adopted. After Santoshchandra Majumdar, Premchand Lal was in charge.
197. P.C. Lal, op. cit., pp. 82-3.
198. Rabindranath Tagore, 'Astha O Babostha', *RR*, op. cit., vol. XII, p. 68.
199. Rabindranath Tagore, *Palliprakriti*, op. cit., pp. 6-7.
200. As his knowledge of the countryside increased, so did his dismay at the apathy of the common rural folk to basic health care. He bemoaned the fact repeatedly in his short stories that give insightful references into the diseases afflicting rural life.
    'Taraprasannar Kirti', 'Ghater Katha', 'Nishithe', 'Drishtidan', 'Sampatti Samarpan', *Galpaguchcha*, Visva-Bharati, Calcutta, 1990, vols. 1-4.
201. The poet divided Birahimpur (Shilaidaha) *pargana* into five *mandals* each under an *adhyaksha* or head. The first five appointed by him were Kalimohan Ghosh, Bhupeshchandra Roy, Anangamohan Chakraborty, Pyarimohan Sengupta and Akshoy Chandra Sen. These *adhyakshas* were to establish village societies, so that the villagers would themselves look after their own welfare, maintain roads, remove water scarcity, settle disputes through arbitration, set-up schools, clear jungles, set-up grain stores to ward off famines. Rabindranath instructed them to

build the houses of tenants and plant banana, pineapple, date palm and other saleable fruit trees. Since he knew that the agrarian village constituted the basis of self-government, he did his best to implement it in practice. Prabhat Kumar Mukhopadhyay, op. cit., vol. 2, p. 187; Rabindranath Tagore, 'Russiar Chithi', 28 September 1930, *RR*, vol. X, pp. 683-4.

202. The poet himself was a keen practitioner of homeopathy. Villagers queuing up for medicines for some common ailment in front of his chamber were a familiar sight. He also had great faith in biochemical medicines and was influenced by the ideas of the German medical practitioner Dr. Susler. His fame spread so fast that even the local English officials came to depend on his medicines for common ailments, viz., sore throat, cholera and fever. In a letter written to his wife the poet informed that the inspector came to him for homeopathic cure to a sore throat. Letter dated January 1890, Shajadpur, *Chithipatra*, Calcutta, 1990, vol. I, p. 1. Ref. Prasad Bhattacharya, *Chikitsabid Rabindranath*, Calcutta, 1890, Introduction.
203. Sachindranath Adhikari, op. cit., p. 72.
204. The poet's disdain for the villagers' blind beliefs in the efficiency of the local *ojha* and hermits was apparent. In his novel *Gora*, the poet illustrates how the local carpenter was prevented by his wife from informing Gora, the protagonist of the novel, when a fatal disease afflicted their son. She feared that Gora would insist on proper medical treatment rather than depend on the local *ojha*. Rabindranath Tagore, *Gora*, *RR*, op. cit., vol. 9, p. 74.
205. There was the case of a villager attacked by a crocodile being kept in the hospital and treated as he could not be shifted. Sachindranath Adhikari, op. cit., pp. 427-9.
206. Fines were collected from rural folk for all misdeeds. This helped to finance health programmes. The poet's trusted lieutenant here, Atul Sen, made rigorous efforts to fight cholera when it spread like an epidemic in 1916.

     To meet expenses, Tagore arranged for the enactment of the play *Phalguni* at his ancestral house in Calcutta. The money collected from the sale of tickets was used for the purpose. Prabhat Kumar Mukhopadhyay, op. cit., vol. 2, p. 436.
207. Rathindranath Tagore, *Father as I Knew Him*, Centenary volume, 1861-1961, Visva-Bharati, Calcutta 1961, p. 6.
208. L.S.S. O'Malley, op. cit.
209. Malaria occurred frequently since 1871. It has been reported that between 1894 and 1929, 2.49 to 19.6 per cent patients died from malaria alone. Hasim Ameer Ali, *Thirty Eight Year's of Rice Yields*, Calcutta, 1934, quoted in Sinha, Dikshit, 'Birbhum Jelar Swasthya O

Swasthya Babostha' in Ajit Mondal (ed.), *Paschimbanga*, Calcutta, February 2006, pp. 75-6.

Smallpox became rampant since 1904. Among other common diseases that afflicted the region were cholera, filaria, *kalazaar*, leprosy, tuberculosis, pneumonia and sexually transmitted diseases. Harry Timbers, 'Report on the Medical Conditions in Birbhum District', *Visva-Bharati Quarterly*, 1930.

210. *Visva-Bharati Bulletin*, no. 25, 1936, p. 3. Rural reconstruction work in Sriniketan began on 6 February 1922 under the guidance of Leonard Elmhirst.
211. The poet regarded malaria as the greatest menace in the region.
212. The poet made the occasion memorable by composing a song *eso eso he trishnar jal* (come come o water to quench thirst). L.K. Elmhirst, op. cit., p. 77.
213. Gopal Chattopadhyay played a commendable role in the eradication of malaria in Bengal. He set-up an anti-malaria society. He influenced Rabindranath who set-up a registered society in Shilaidaha to fight the deadly disease. Later Chattopadhyay also helped Elmhirst at Sriniketan. Sachindranath Adhikari, op. cit., p. 31.
214. *Visva-Bharati Bulletin*, no. 25, 1936, p. 3.
215. Rabindranath Tagore, 'Malaria' and 'Samavaye Malaria Nibaran', *Palliprakriti*, op. cit., pp. 123-40.
216. Bipasha Raha, 'Rabindranath and the Cooperative Idea', *Bengal Past and Present*, vol. 119, 2000, p. 12.
217. *Visva-Bharati Bulletin*, no. 25, 1936, p. 13.
218. Sasadhar Sinha, *Social Thinking of Rabindranath Tagore*, Asia Publishing House, London, 1962, p. 104.
219. Sugata Das Gupta, op. cit., p. 62.
220. Ibid.
221. The pond in front of the local Gropinath temple was cleaned.
222. Sachindranath Adhikari, op. cit., pp. 32-3.
223. Sukumar Mullick, *Rabindranather Pallichinta* (Bengali), Nabapatra Prakashan, Calcutta, 1989, p. 107.
224. Prabhat Kumar Mukhopadhyay, *Santiniketan Visva-Bharati*, Visva-Bharati, Calcutta, 1961, vol. 1, p. 121.
225. L.K. Elmhirst, op. cit., p. 77.
226. P.C. Lal, *Reconstruction and Education in India*, 1966, p. 92.
227. *Visva-Bharati Bulletin*, no. 25, July 1936, p. 14.
228. Every year repairing of roads and drains and cleaning of tanks were undertaken twice according to the need. The work was done with the help of the young men of the village enrolled as *Bratibalaks*.
229. Bhubanchandra Sinha, zamindar of Raipur built it.
230. Prabhat Kumar Mukhopadhyay, *Rabindrajibani*, op. cit., vol. 4, p. 61.

231. The poet expressed his opinion in the matter in a letter written on 20 September 1925 to Madam Sangar. Ibid., p. 46.
232. Somendranath Basu, 'Rabindranath O Paribar Parikalpana', *Ananda Bazar Patrika*, 16 Baisakh, 1381 B.S.
233. Popularly known as A.E., George Russell's *The National Being* greatly impacted upon the poet's thoughts. The poet was also fortunate to count among his friends eminent scientists like Acharya Jagadish Chandra Bose, with whom he shared his thoughts and findings on vital matters related to the work.
234. 'Peace, Water and Toilets', *The Telegraph,* 11 November 2006, vol. XXIV, no. 126.
235. The poet left it to Elmhirst and his team at Sriniketan to break the clod and upturn the soil. Krishna Kripalini, 'Foreword', in L.K. Elmhirst, op. cit.

CHAPTER 5

Changing Images of the Rural World and the Political Programme

Sholosho chashar adh chaatak buddhi
Tao aa jalkhabar bela parjanto[1]

The hapless peasantry continued to be a source of concern to the informed intelligentsia. However, the question of rural reconstruction never really became fundamental to the search for certain types of nationalist alternatives to colonial institutions by the dominant section of the elite group. Nevertheless, it could not be ignored in the political process, in the nationalist attempt to acquire a broader base for the anti-imperialist struggle. Rural issues were harnessed by an active intelligentsia to ponder and debate on penetrating questions of probable links between the extent of constructive work in the countryside and mass mobilization. With the beginning of the *Swadeshi* movement and the emergence of extremism as a political force, tentative attempts were initiated to bring the rural masses within its fold. Mainstream Indian nationalism in Bengal in early twentieth century tried to accommodate the self-strengthening movement or *atmashakti* that Rabindranath, the main ideologue of a non-political constructive programme that relegated the political agitation to a secondary position, advocated. However, he was able to motivate attempts by only a section of the literate to integrate the nationalist cause with the socio-economic demands and aspirations of the common people in the rural areas. Even though for Tagore rural reconstruction work was a goal in itself and not an instrument for mass mobilization, indigenous attempts in the twentieth century were motivated primarily by the need to mobilize the rural masses for the nationalist movement. It was perceived as a response of the Indian people to colonialism, a diverse technique to be adopted during the greater struggle. A new trend that emphasized on self-reliance, village level organization and constructive programmes, soon informed intellectual activities in phases. Intellectual pre-occupation with the issue was not totally unconnected with the general *political* ambience.

Two distinct trends could be discerned in their handling of such matters. At one level, agrarian thinking was increasingly a part of organized politics. Political parties and organized politics were a mode of functioning of the literate under certain historical circumstances. The first quarter of the century was a period of intense political activities. It went through multiple phases beginning with the anti-partition struggle and including Congress and Swarajya politics, the spread of the *Praja* and the communist movements and the prolonged controversy over the amendment of Bengal Tenancy Act during which time those with political leanings addressed selected agrarian issues. The legislature and public meetings as also the press served as forums for expression of convictions and allegiances. Agrarian relations were the focal point of interest. At another level there was also increasing literary representation by those not directly involved with politics. They belonged to diverse occupations from litterateurs to professionals. Some like Nagendranath Gangopadhyay were active workers in agrarian experiments. The literati's perception of rural life was completely recast by a process set in motion by developments initiated earlier. The impact was such that it imprinted on the thought, literature and political process of the period and gave it a new dimension.

LITERARY REPRESENTATION

There was an outpouring of creative writing, using different literary genres in Bengal in the first quarter of twentieth century. The peasant and rural society continued to occupy centre-stage in both Hindu and Muslim literature. Contact with the West left a deep imprint on the literate. Familiarity with Western thought gave a new dimension to their thought process, which was reflected in their work. Since the First World War, the impact of socialist thinking on literature may be discerned. The gradually worsening economic situation explains this socialist leaning. While agriculture languished, unrest became a normal feature in industries. Repressive measures of imperial rulers and economic degradation had an alienating impact. With increasing strain, resentment and restlessness deepened. It was in this state of things that the Bengali literary mind was attracted to socialist thinking that had become a force in Europe. Of all European developments in contemporary times, the Russian Revolution impressed the Bengali youth most. There developed a new attraction for Marxist thinking and Russian literature since the 1920s. The popularity of Russian literature through translations fanned socialist leanings. Literary minds had long ceased to be satisfied with the life, achievements, sorrows and anxieties of social superiors alone. While no attempt was initiated to formulate a new ideology or preach a change of the base of our society, a new awareness of socio-economic maladies was prompted. Therefore, what actually happened was that attention was diverted to the downtrodden social groups after the pattern of the socialists. The glaring inequalities in society became the main target in many of these works. The dreams and depressions of the agriculturists and labourers became the theme of many a social novel. At the same time, a feeling of antagonism against the aristocracy was visible, particularly in the writings of the Muslim litterateurs.

Rabindranath at the crossroads of two centuries continued to portray a vibrant image of rural society and inspired many of his younger contemporaries. Literature reflects the literati's attempts to understand agrarian society. In the event, contemporary literature, influenced by a spirit of humanism acquired a progressive trend. In their handling of issues related to the peasantry and rural society, creative writers displayed a new level of maturity. In Muslim literature, the dominant themes were exploitation of ryots by zamindars and moneylenders and the formers' miserable plight.

Ryots were portrayed as ill-fed, ill-clothed beasts of burden. By the 1920s, antagonism against Hindu exploiters became pronounced. Abul Hussain, an eminent and popular litterateur who enjoyed a considerable readership, in his quest for the cause of poverty in a country rich in natural resources underscored in powerful prose the machinations of a parasitical class of exploiters.[2] In his novel *Durbipak* Nazirul Islam Sufian, a writer and educationist, described in powerful prose the kinds of atrocities committed by Hindu zamindars.[3] The plot revolves round Yaqub, a small peasant whose rent was remitted at a time of crop failure by the zamindar only when his mother gave her seven-year daughter to the latter. The girl later ended up in Calcutta. Syed Ismail Hussain Siraji in the 1920s even hinted at an emerging Hindu marwari nexus.[4] He was apprehensive of the fact that marwaris were increasingly investing commercial profits in *jot*s and zamindaries. Their usury operations had already entrenched them in a position by which they could exercise control over the Muslim peasantry through the mechanism of debt. If such situations persisted for a couple of decades more, a land owning peasantry would be relegated to memory and the Muslim community would comprise only workers, menials and servants.[5]

Perhaps the most lucid tales of oppression were told by Kazi Nazrul Islam, one the greatest literary figures of the twentieth century, in some of his early writings.[6] His poetry and essays give graphic descriptions. In the poem 'Gariber Byatha', he bemoaned that the hapless peasant families who grew the golden corn but had to depend on the charity of the rich. They were denied proper sustenance and any sort of medical help.[7] 'Chashi' expresses his anguish for those whose hard labour provided the rich with all amenities of a luxurious living.[8] In 'Chor dakat'[9] in the 1920s the king, the rich, the mill owners and zamindars were represented as robbers more despicable than those whom society recognized as thieves and robbers, as they robbed the poor of all that was rightfully theirs.[10]

These men were equally vocal in their denouncement of rural moneylenders. Muhammad Maniruzzaman Islamabadi, a writer, journalist and political activist who was associated with the *Krishak Praja* movement warned peasants in the 1920s against Hindu moneylenders and their own expensive habits. The propensity of Muslim peasants to get embroiled in prolonged and avoidable legal suits and incur heavy debts due to conspicuous expenditure, viz., weddings, was well documented in his essays.[11] The peasant was depicted as a pitiable figure always in debt whose property was sold for arrears to Hindu creditors. Consequently, the picture of Hindu settlements coming up in Muslim villages became a frequent occurrence. He suggested rigid control over the social life of Muslim peasants to protect them from themselves. The suicide of Sadeq, protagonist of noted educationist Kazi Imdadul Hoque's novel *Abdullah*, when his father Madan Gaji lost even the land on which stood his homestead, to the moneylender Digambar Ghosh for arrears of interest on loan taken during the former's marriage was an attempt at a realistic pen-portrait of the life of an average Muslim peasant.[12] Hoque was particularly vocal against the moneylending system.

Abul Hussain criticized Muslim *maulvi*s for issuing *fatwa*s stating that Islam forbade usurious interest.[13] They offered no alternatives. Nevertheless, the poor Muslims, he explained, needed to be protected from moneylenders and the vicious cycle of debt and interest payment. Otherwise, the chances were that either Muslims would be starved to death or they would take recourse to retaliation through revolts.

As a remedy, Muhammad Abdur Rashid in *Banga-Moslemer Durabasthya O Tahar Pratikar* proposed banks in each village to facilitate borrowing at low rates. Besides, arrangements could also be made to organize funds in every village, from which the incumbents could get access to soft loans.[14]

The peasant in Abul Hussain's *Banglar Balsi* in 1328 B.S. was an easy prey to the tall promises of Western type democracy made by their immediate social superiors who could never identify themselves with the protracted struggle of the former for survival and equal rights.[15] Their interests were diametrically opposite. The Bengal peasantry was always ill-fed, ill-clothed and permanently in debt. Hussain was convinced that if such conditions persisted there would be an upheaval before long and was rather sceptical about the success of a united political struggle that would serve the interests of all social classes alike. The threat of a revolution would not really disappear until such time when zamindars undertook agrarian improvements and ryots became free in every sense of the term.[16] He warned zamindars about the possible impact of Bolshevism. These hints at peasant revolts at the height of the nationalist movement are significant. The political struggle as also the Bolshevik Revolution informed his understanding of the social crisis.

Looking for means to solve the problem of economic backwardness of the peasantry, Islamabadi suggested encouragement of trade and commerce and reduction of wasteful expenditure. He was convinced that the zamindari system should be abolished. However, as this was both a difficult and time-consuming process, other immediate measures need be adopted, viz., launching of a movement for reduction of rent, cancellation of *abwabs*, waiving of rent during natural calamities, fixing the minimum price of jute, starting agricultural banks, exempting ryot-tenants from *chowkidari* tax and, construction of shallow canals. Inspired by the ideas of Friedrich List,[17] Abul Hussain advocated promotion of all types of indigenous artisan industries necessary for production of commodities indispensable for survival to facilitate attainment of economic self-sufficiency.[18] He posited that in the modern world, mill owners and machines had killed the 'inner man' inside the workers.[19] It was the struggle of the 'inner self' of the latter, he observed significantly, to exert itself that found expression in 'socialism', 'collectivism' and 'Bolshevism'. Industrialization influenced rural life while urbanization gradually destroyed the villages, which constituted the nucleus of Indian civilization. Hussain explained that participation of the educated youth, with their knowledge of scientific techniques of production and trade, in agricultural production could help in the process of economic revival. He also advocated the development of pisciculture as a source of income for the rural population. There was a great demand for fishes in Bengal and it could become an important item of trade. Income could be supplemented by engaging in cattle rearing and poultry farming. The educated young population particularly in the countryside would be better advised to undertake such activities instead of competing with the better-educated Hindus for the few job opportunities that were available. Hussain posits unequal distribution of wealth for the economic distress of the rural masses. He appealed to the educated and prosperous Hindus to help in the economic recovery of the poverty-stricken Muslim masses. He shared with Rabindranath the same faith in the potential of the cooperative principle. His suggestion of a socialistic economy as the only viable solution to the crisis posed by capitalism and imperialism is significant. By this, he meant a classless society where factors of production were commonly owned and there was a redistribution of wealth intended to attain some parity in income.

Other piecemeal suggestions were made. Siraji advised Muslim peasants engaged in jute production to unite and establish monopolistic control over the jute trade that was hitherto dominated by English and Hindu-marwari traders. The English in particular, he insisted should not be allowed to participate and the Muslims should wrest control in the foreign market. Thus, the Muslims peasants so long denied a share of the profits would benefit.[20] He also suggested in order to promote Muslim participation in all kinds of commercial activities, business enterprises with initial working capital varying between Rs. 5 to 10 lakh be opened in each commercial centre and district headquarters. To enable the poor to invest and be active partners, the value of shares should be between Rs. 5 and 10.[21] Sadat Ali Akhand who advised the Muslim youth without access to capital to undertake insurance companies[22] shared Siraji's belief in the importance of such enterprises. The Hindus, he observed, had been able to take advantage of the reluctance of their Muslim counterparts towards such activities and prosper.

Review of Muslim literature produced by the newly emergent literati underscores significant trends that marked their mentality. Their agrarian perceptions did not remain separated over the years from their political beliefs and became gradually obscured as some sort of duality in their attitudes and relationship with the complex reality of peasant life becomes visible. In tune with the Muslim quest for identity and influenced by the attempts at mobilization of Muslim peasantry, particularly in the eastern Bengal districts, they were concerned mainly with the fates of Muslim zamindars and peasants. Where their concern was the former, they overall blamed Hindu officials in zamindari estates for ruining their masters. Except in the early works of Masarraf Hussain, the oppressors of Muslim tenantry were usually Hindu zamindars and moneylenders. They ignored the fact that Hindu peasantry fared no better. Nor were Muslim zamindars and officials less oppressive in their relationship with co-religionists.

Yet, simultaneously many of them understood that without the active support of the economically prosperous Hindus, the backwardness of the Muslims would continue to persist. Therefore, they often appealed to Hindu merchants and traders to involve Muslims in cooperative economic ventures. Most of them did not promote communalism. Siraji for instance was an active political activist loyal to Surendranath Banerjee. He was active in the anti-partition struggle in 1905-6 and supported the Boycott and *Swadeshi* movement that was resisted by a large section of the lower caste Hindu and Muslim peasantry. They regarded it yet another instance of coercion by Hindu zamindars and the elite politicians. They found their own methods of resistance. Siraji however, was pan-Islamist and anti-British in his perceptions. He was imprisoned in 1909 for his political bearings and his book *Anal Prabaha* was banned. He was a member of the Indian National Congress.[23] Maniruzzaman Islamabadi too was active in politics and participated in the Khilafat movement. As a member of the Indian National Congress, he later joined in the Civil Disobedience movement and then in the 1930s became associated with the *Krishak Praja* movement. He insisted on promoting awareness among the peasantry and participation of the educated youth in their economic betterment. Neither he nor his contemporary Abul Hussain believed that Muslims would benefit from adopting the policy of separatism. In fact, they represented the new line of political thinking promoted by the young generation of Muslim leaders who had no faith in the pro-British anti-Hindu policy, so far the dominant trend among the old guards closely associated with the Aligarh school. They felt it imperative to break the stranglehold of foreign traders on the indigenous economy as a pre-condition for balanced economic growth. They warned against frequent recourse to the law courts, as it benefited none but the foreign masters and advised settlement of disputes through arbitration.

Muhammad Lutfar Rehman, author of the considerably popular novel *Raihan*, in fact, stressed the importance of self-dependence.[24] It was published at a time when the Khilafat movement was just gathering momentum and thus helps us to contextualize the author's ideas. He denounced import of foreign goods and advocated self-production of necessities. Writer Abul Mansur Ahmed emphasized the importance of boycott and use of indigenous *khaddar*. This alone he felt, was the only way to solve the problem of food scarcity and poverty.[25] Herein lay the road to economic recovery.

Agrarian thinking of these men revealed two angles in their handling of crucial issues. The first was a focus on zamindari and moneylending operations and the other was an attempt to locate the peasant and the rural world within the existing overall economic system. There was increasing tendency to re-examine theological dictates as possible causes for Muslim backwardness. All kinds of wasteful socio-religious expenditure were frowned upon as roots of burden on an already overburdened peasantry. Abdur Rashid questioned the legality of the *fatwa* issued by the *maulvis* stating that Islam forbade charging of interest. Criticizing the dictum, he said that the Koran prohibited *reba* or interest on interest. Nevertheless, generally interest was not forbidden. Besides, Islamic law stated that interest was not illegal in *dar-ul-harb*.[26] It made no sense to him that poor Muslim peasants bled paying interest to Hindu moneylenders. There was also a general dissatisfaction with colonialism and capitalism. A socialistic type of economy seemed to some an alternative solution to the crisis posed by imperialism and colonialism.

However, most of their works were not essentially anti-colonial discourses and did not question British rule. The only probable exception was Nazrul. He raised his voice against the exploitation of the weak by the strong in a poem titled 'Fariad'.[27] He preached revolt and the ultimate victory of the weak and exploited.[28] In a poem 'Daridra more paramatiya' he expressed his desire to arrange for education, medical facilities, food and clothing for the poor and the needy.

Mention must be made here of two remarkable essays written by him early in his career—'Mandir O Masjid' and 'Hindu Mussalman'.[29]

Published in the *Ganabani*, in the early 1920s the mouthpiece of the communists in Bengal, these essays written at the time when the country saw communal riots in Kanpur, Calcutta and other areas, are remarkable in that they speak of the author's faith in communal harmony. He blamed both Hindu and Muslim fundamentalists for such mindless violence. In his understanding, the common person—both Hindus and Muslims—was the ultimate sufferers. Nazrul was a confidante of Muzaffar Ahmad and actively associated with the Workers and Peasants Party. He sought to inspire both with the message of truth and justice and repeatedly asked them to unite for their rights. Through the power of revolution, he hoped, all injustice in society would cease to exist. The establishment of a socialistic economy he hoped would further the cause of an exploitation-free society.

The Muslim intelligentsia, however, failed overall to provide a plan for agrarian recovery. They did not realize that without this there could be no real improvement in the condition of the peasants. Their agrarian perceptions were primarily limited to highlighting the miserable conditions of Muslim peasantry and seeking reasons for it. They dealt with issues, viz., agrarian relations, role of moneylenders in the rural economy, existing class antagonism and the exploitation of the upper classes. Nevertheless, they ignored one vital aspect of the agrarian economy. They did not handle issues related specifically to agriculture on which the whole rural structure was based. Vital issues, viz., agrarian yield, tools and techniques of production, adoption of scientific methods of farming and such others were overlooked. Most of them failed to suggest any positive steps as to how agricultural production could be improved. They overlooked age-old techniques of agrarian production practised by Bengal peasants. There seemed to be a general lack of awareness about how agriculture had been revolutionized elsewhere in the world, particularly in Europe. Therefore, they could not suggest ways and means for adoption of Western scientific methods here to combat agrarian backwardness.

This shortcoming may have stemmed from the fact that most of them lacked any sort of practical experiences of agriculture. They had little first-hand knowledge of agriculture. That is probably why they failed to appreciate that without changes in agrarian techniques, increase in productivity, provision for easy credit and better system of agricultural marketing economic prosperity of the peasantry would continue to remain a distant dream. These Muslim men of letters made no attempt in their literary creations to distinguish between different classes of peasants and preferred to consider it as a category with interests diametrically opposite to that of the landholding section, forgetting the fact that there were exploiters even within the exploited group *vis-à-vis* their social inferiors. These litterateurs failed to relate issues vital to the peasantry with greater rural problems and ignored the importance of rural development as a whole. Nevertheless, the issue did not disappear from public discourse.

CONSTRUCTIVE ATTEMPTS

A section of the Hindu intelligentsia, not actively political, continued to be sensitized to the urgent need for rural revival. Away from the demands of a hectic political life, they were inspired by the idea of redeeming the neglected village. Young, Western-educated and informed by the spirit of earnest activism in an age which is characterized as a 'hiatus between the myth of renaissance improvement and national deliverance'[30] they reflected the mood of their period. They belonged to the particular social environment characteristic of upper middle class intelligentsia culture in early twentieth-century Bengal. While their vision was rooted in a genuine urge to fulfil the needs of the village, they also envisaged constructive work in the village as a positive means for popularizing nationalist ideas. In the first quarter of the century, the question of amending particular sections of the Bengal Tenancy Act also became crucial.[31] The need for changes was admitted by all sections of the populace. The government also expressed its desire to make necessary changes. This once again reopened a public debate on rights of the peasantry and focused intelligentsia attention of the intelligentsia on rural society. Among the most hotly debated issues were the questions of granting the right of free transferability to occupancy peasants and giving occupancy rights to under-ryots and *bargadars*. Not all these men were creative authors. They came from diverse backgrounds. Nagendranath Gangopadhyay, Radhakamal Mukhopadhyay, Srinath Dutta, Nityagopal Mukhopadhyay and Dwijadas Dutta among others were the best representatives of this composite group, which was not totally untouched by the general political ambience.

The nationalist quest for *Swaraj* informed their thinking in the first two decades of the twentieth century. Nevertheless, they were convinced that *Swaraj* had a far-greater connotation than just a political goal. *Swaraj* in its real form could only be attained if the needs of the village were fulfilled.[32] Nagendranath Gangopadhyay, however, admitted that times of heightened political activities were best periods for concentration on work for rural rejuvenation.[33] He was convinced that political leaders—extremists, moderates and all others—could use rural developmental work as a technique for mobilization of the rural masses for the political struggle. Even though he understood that such work should best be left out of the purview of politics, he admitted that under the circumstances of colonial rule and its conflict with anti-imperial forces, such a contingency could not be avoided. In that case, the general awakening and the nationalist aspirations of the educated urban population could be channellized for constructive work in the rural areas, which would in turn reduce the sense of alienation of the rural masses from the mainstream movement. First Rabindranath and then Gandhi's clarion call for devotion to such work, though they were worlds apart in their methods and techniques, their focus on the miserable plight of the peasantry, poverty and absence of education that prevailed in the countryside gave a fillip to such endeavours. Nagendranath associated such work as part of the mighty labour of nation building.[34]

Arguing from the point of view of an economist, Radhakamal Mukhopadhyay advocated it for healthy and balanced growth of the country.[35] While Nagendranath warning of the futility of disjointed efforts of *Swadeshi* leaders advocated a concerted plan of action, a pragmatic activist Radhakamal was aware of the difficulties of undertaking such work in a colonial country. An antagonistic government could thwart all attempts. He was in favour of instilling a spirit of self-reliance in villagers. As an economist, he was convinced that adoption of a comprehensive programme of village reconstruction alone could initiate material improvement in the condition of ryots, the single largest social category. He was, however, sceptical about the viability of isolated efforts[36] at social welfare in securing permanent gains.

Attempting to get to the root of the ills besetting the countryside Radhakamal found it in the increasing migration of the potentially rich and educated population to the urban regions. This deprived villages of all their youth and vigour. He also had a profound grasp of land as a site of social conflicts, struggle for power and agrarian tensions. Agrarian tension, he argued, was so pervasive that it shaped the psyche of almost all villagers. Constant and expensive litigation over land constituted a heavy drain of peasant resources and further aggravated the existing disharmony. Peasants were also unaware of the use of improved qualities of seeds and fertilizers. Even when these were made available to them, they were unaware of their exact price and thus made to pay more. Denied of crucial investments and much needed changes in techniques, agriculture languished. The law of diminishing returns informed agricultural production. Radhakamal was severely critical of the growing trade in grains at times when most villages were facing acute shortages. Peasants were denied any share in the profits. Most interior villages were still denied any access to basic educational opportunities and amenities for health. This analysis of the condition of ryots did not in any way vary from that of the nineteenth-century intellectuals. It indicates the unchanging picture of colonial rural India.

In contemporary journals, the work-plan was hotly debated. Ramananda Chattopadhyay, the doyen of Indian journalism, gave the impetus to such intellectual discourses. The erudite editor of *Prabasi* and *Modern Review* in early twentieth century gathered round him a group of enthusiastic, highly educated and well-informed young men, who often excelled in their chosen areas. Chattopadhyay gave them the forum to express themselves on wide-ranging issues related to life in a colonial state.[37] They debated and discussed within its parametric constraints. Chattopadhyay a close friend of Rabindranath was fired by reformist zeal. Apart from social regeneration, he encouraged the quest for rural uplift. He agreed when Kali Kumar Mitra asserted[38] that in this difficult mission, a rural reformer needs to remain undaunted. Since the work was a long-drawn process and difficult, Nagendranath suggested that it was imperative to have as volunteers men of strength and character, with a firm conviction in the eventual success of their work. His long stay in the countryside and his active involvement in such work had brought home the fact that, it would be an uphill task to overcome the belief of the elderly rural population in the ultimate futility of such work and make them active partners at least in the early stages. The rural youth was expected to play the lead role. Nagendranath articulated in powerful prose that for rural reconstruction work to attain any degree of success it must cater to the needs of all sections of society irrespective of caste and religion. He was in particular concerned about the untouchables and other downtrodden strata of society. He was emphatic that all caste distinctions in Hindu society should be obliterated. He focused on the agricultural class, the *paria*, as the largest class within society, paralyzed by poverty and lack of education and whose interests were ignored so far by the upper class Hindu society. All this he felt, was in greater interest of the nation and a pre-condition to political *Swaraj*. In his reckoning, social bondage was more oppressive than political. So constructive work should begin with those occupying the bottom layers of society—peasants and low caste groups, viz., hadi, bagdi, dom, chamar, etc.

As to the *modus operandi* that was suggested, the first task was to divide workers and conduct survey work. Extensive knowledge of the countryside was an essential prerequisite for work. All necessary information needed to be accumulated about individual villages before remedial work could be started. Nagendranath cited instances of how some zamindars, inspired by the idea of *Swadeshi*, had appointed volunteers to undertake beneficial work in their estates. However, all these efforts had ultimately failed, primarily because workers lacked relevant information about individual needs of villages within the estates. Education was to be given top priority. Educated members of lower classes were to be inducted into the working committee of primary schools. Kali Kumar even suggested that in every school a girls' section be opened to encourage the lower classes.[39] Opportunities for higher education was to be provided to deserving rural children.

Unless the people were inspired enough to help themselves, all development would leave their souls untouched.[40] No external obstacle, argued Nagendranath, was as powerful as internal ones. The actual obstacles existed in the minds of the people—their lack of belief and confidence in themselves and their inertia that prevented them from attempting to improve their lot. What was lacking in the villages was 'spirit of the community'. To foster this it was imperative to focus on the mental health of villagers. In this respect, he felt, the efforts of external agencies, viz., district boards, magistrates or the police could not go much further. Nor could the establishment of *sabhas* and *samitis* in villages, set up to look after local needs, help much.

The foremost task advised, Radhakamal, was to train villagers in the virtues of self-reliance and motivate all those who were willing to work.[41] Cultivators and rural artisans should be made self-sufficient so that they could undertake, without any expectation of government help, diverse activities to suit their needs. Radhakamal was firmly convinced that the active participation of the urban middle class, educated and prosperous, in agricultural, trade and business was vital for the material advancement of the rural poor. This he felt would foster cultural development and independence of thought among villagers and accelerate the process of nation building. Once initiated, the pace of rejuvenation work, started in as many villages as was possible simultaneously, should be slow and cater to the basic needs of rural life in general.

Introduction of the cooperation system at all levels of rural work was the guiding principle advocated by these men. It is significant that they were well acquainted with the Irish cooperative movement and the ideas of Russell and Plunkett.[42] Cooperative societies, they suggested should co-ordinate activities vital to rural life. Since historical times, argued Nagendranath, cooperation had been the way of life in rural society. Collection efforts underlay many activities in Bengal's rural life, viz., community workshops and rituals, moneylending, buying and selling of cattle, fairs and festivals. However, years of colonial domination had destroyed those human qualities, viz., mutual trust and self-esteem, which made cooperation especially in enterprises related to production and consumption difficult. Success of rural reconstruction work depended on creating an awareness of the advantages of cooperative enterprise. Nagendranath advocated its application to all types of activities, viz., agriculture, poultry work, dairy farming, rural industry, etc., to make villages self-sufficient and remove poverty, the bane of rural life. Radhakamal in particular was inspired by the success of the cooperative moment in Germany, Holland and Ireland. He observed that while in India the cooperative moment, which was still in its infancy[43] was primarily restricted to providing soft loans to rural producers in European countries it was responsible for integrated welfare activities, viz., making available to producers high quality seeds, fertilizers and necessary agricultural implements.

Nagendranath argued in favour of the *mandali* system, outlined by Tagore in the 1890s, to encourage self-sufficiency and joint responsibility for education, health and finance.[44] When he was talking of rural development on these lines, he was actually echoing the Tagorean concept of *atmashakti*.[45] Nagendranath was categorical in his denouncement of the claim that multifarious problems facing the country could be solved through any political reforms of the Morley-

Minto (1909) or Montagu-Chelmsford (1919) kind. The time had come to establish before the country the real ideal of freedom. Justifying the need for promoting self-sufficiency in villages another contemporary Srinath Dutta expressed fear that by the next census, population of Bengal was likely to go up to nearly 5 crore.[46] This would make imperative active participation of villagers in activities related to their well-being. He suggested that *chowkidari panchayat* and the village union was included in each village committee. Those villagers who paid road cess, *chowkidari* cess and all other cesses should elect members of the village committee.[47] These members should be given the power to elect members of an organizing *sabha*. Each subdivision, he suggested, should appoint 1,000 *mandals* for 2 to 3 years. They were to undertake all work of the village committee and appoint members of the organizing committee. They would supervise provision of drinking water, schools, health centres, clearing of jungles, and building of roads and embankments. The money collected in the form of road and other cesses, he said, rightfully belonged to the villagers. He wanted this money to be divided between the village committee, the circle or local board and the district board with a minimum of two-thirds of the road cess being reserved for the former. He wanted this to be made legal. This village committee of manageable size was to be entrusted with the task of maintaining Hindu-Muslim amity. He proposed 40 per cent reservation for Muslims in all village level organizations.[48] Another rural enthusiast, Amarnath Dutta, advised focus on provision of job for dwellers particularly in villages situated away from towns and railway tracks.[49] This would check migration of the semi-educated to towns. He suggested shifting of some branches of judicial, revenue and administrative departments at low costs to the villages. Besides development of communication system could help villages merge as commercial linkages.

All work was to start with the peasants and the artisans who occupied the lowest strata in rural society. Radhakamal advised organizing in each village a common store or *bhandar*. The village *panchayat* or *mandali*, entrusted with the task of monitoring, was to remain in charge. Funds mobilized within the village would be deposited and the *panchayat* would acquire the basic requirements, viz., clothing, sugar, salt, etc., at minimum price. These were to be made available to villagers at wholesale price from the *bhandar*. Radhakamal argued that if the sales of the *bhandar* revealed an upward trend then, the *panchayat* instead of acquiring these goods from neighbouring towns and markets could themselves employ artisans to produce those at flat rates. This would give the sorely needed fillip to rural industries. The village weavers and smiths would also be motivated to send their products to the *bhandar* and from there acquire their grains in exchange.

Agriculture being a crucial component of any rural revival programme these young literates since 1910 argued in favour of modernization of agrarian techniques and introduction of scientific agriculture on lines similar to that advocated by their more famous senior contemporary. Nagendranath's arguments were based on the premise that without improvement in the quality of agriculture, there could be no real uplift in the living standards of the peasantry. For improvement in the quality of crops produced, he suggested, imparting of the knowledge of the use of fertilizers and good quality seeds to farmers. Working in cooperation, peasants could pool in their resources and invest in procurement of modern machines. Collection and sale of crops could also be done on a cooperative basis. A cooperative grain store could be opened in each village to solve the problem of periodic food shortages. The system of agrarian marketing could be modernized and thus ensure higher returns for the peasantry. Explaining how this could be done, Radhakamal stated emphatically, peasants should stop selling their products to brokers and wholesalers who denied them a fair price.[50] He proposed that they should arrange on a cooperative basis for the sale of their produce at wholesale price that would thereby enable them to control unrestricted export of crops. Sale of artisan products abroad could also be organized in a similar manner. He also proposed exhibitions and competitions for agrarian and artisan products in order to generate a demand for them in *mofussil* and urban areas.

The literati were also in favour of new technical knowledge being imparted to peasants. However, at a time when most were in favour of the educated urban youth taking an active interest in agriculture there was Nityagopal Mukhopadhyay in 1902-3 striking a discordant note.[51] He warned those attempting agricultural experiments without experience of grave financial losses.[52] A better and more viable option, he reasoned, was for the *bhadraloks* to carry on such work based on mutual sharing with the peasants. The former would be better advised to provide land, good quality seeds and specific implements, while the latter would provide farm animals and labour.[53] This would benefit both. Nityagopal also advised those interested in a career in agriculture to acquire formal training and was in favour of provisions for such education from the primary school level.[54] As to how peasants could be given the new knowledge he was all for agrarian education being imparted in village schools. A contemporary, Dwijadas Dutta[55] was in favour of peasants being trained to classify land in Bengal depending on their quality. He observed that they should learn to distinguish between lands suitable for grains from that suitable for fodder crops, for pastoral farming, or for cultivation of fishes or for that of bees, lac and silk.[56] He advocated mixed farming albeit on a small scale and consolidation of land holdings. He advised peasants to use agricultural waste as fodder for farm animals while animal wastes could be used as fertilizers and practice crop rotation. Overall, he preferred general agriculture to remain in the hands of peasants while interested *bhadraloks* would find cattle farming and cattle trade more profitable. He found in the region great demand for dairy products.[57] In fact, Radhakamal[58] suggested that a general dairy be set up in each village where there would be scope for production of pure milk and other milk products on a cooperative basis.[59]

The problem of rural credit was given priority by these young intellectuals. Reports of moneylending oppression informed their understanding. They regarded it as a serious hindrance to agrarian recovery even while appreciative of the credit needs of peasants. Instead of peasants taking loans individually from local moneylenders, grain dealers or moneylending zamindars at high rates of interest, the economist in Radhakamal suggested collective farming. He proposed that resident peasants of each village should combine, take responsibility for each other's debt and mobilize resources for launching a cooperative credit society. Each peasant was to be made guarantor of the loan advanced to his neighbour. Consequently, this would generate active interest in each other's agricultural pursuits. This would serve as restraint on misuse of borrowed capital and ensure productive use of resources. Once this was done, it would make possible 'collective responsibility'. He cited instances of agencies giving cash advances to peasants, set-up in scattered villages at government initiative, to emphasize the growing realization about the benefits of such enterprises.[60] However, he warned that along with easy credit facilities these agencies should ensure that peasants could modernize their techniques of production and sell their products at a fair price. This alone could accrue permanent benefits to rural society. As to the composition of cooperative credit societies to be set up in each village, Kali Kumar Mitra proposed that a minimum of 15 persons irrespective of caste should combine to form it.[61] The society was to be registered under the registrar of the Bengal Co-Operative Society. After registration, he proposed that the society could appeal for inclusion within the district central cooperative bank. If this appeal were accepted then it would have access to the bank's shares. This would enable the village society to get loans at low rates of interest from the bank. This benefit would percolate down to the villagers. Membership of the society was to be made mandatory for all villages.

The literati were justly concerned about the growing indebtedness among the peasantry. Nagendranath observed how the steep rental and rising costs due to war and other dislocations left actual cultivators with unpaid debts that kept on accumulating, forcing more distress sale of produce. They hoped that imparting to them knowledge of the virtues of thrift, working in union and habit of saving as also providing them with soft loans and facilities for proper marketing would improve the situation.

Another source of concern was the all-pervasive class tension and zamindari oppression in the countryside.[62] Nagendranath blamed the high rental demand and absenteeism for continuing disharmony between zamindars and ryots.[63] Nagendranath held firmly to the belief that, while landlords could not be expected to give up their claims on the rent, they could be made morally responsible for the welfare of tenants within their respective zamindaries and protect them from the local exploiters, viz., estate officials, village constable, moneylenders, etc. Regular visits to their estates and an interest in their zamindari affairs to begin with, could gradually foster better relations between zamindars and tenants and convince the former about the necessity of initiating reconstruction work. Nagendranath suggested that at least for a few months every year, the zamindars should stay in their zamindaries. In these suggestions, his underlying conviction in the innate goodness of zamindars is apparent. He argued that property owners in Bengal who imitated their English counterparts in all respects should also adopt their enterprising attitude towards agricultural practices. He advised them to provide improved quality seeds, fertilizers, and agricultural implements and if necessary, employ agricultural experts. He hoped that they would provide the same support as the government in the USA did to the peasants. He cited instances of how in America the government made commendable efforts to reclaim fallow lands. Its agricultural department had set-up a society and entrusted it with the duty of reclaiming infertile land, guiding peasants on the types of crops suitable for their kind of soil and how best to produce them. Nagendranath desired that the prosperous proprietors too set up similar departments in their estates to undertake such work. This would ensure all-round agrarian development in the villages.

Nagendranath also realized how desperate the land problem was. Therefore, he acknowledged that to solve the problem and to ensure real material benefits to Bengal's agrarian society, an ultimate change in property relations was imperative. Arguing in the context of the public debate on the probable amendments in Bengal Tenancy Act, he suggested that peasants be given the rights to fixity of tenure, fixity of rent and free right of sale or transfer of their holdings. He held that, the right of land should belong to the peasant as the natural tiller of the soil. He agreed with Pramatha Chaudhuri that the Bengal ryot should in actuality be made a peasant 'proprietor'.[64] He did not agree with his mentor that ryots under the jurisdiction of zamindars were better off than they would be if land was transferred to moneylenders. His experiences of the hapless condition of the peasants convinced him of the necessity of ensuring security to the ryots. Unless the peasant was secure in his holding, all efforts to revitalize the villages would be an exercise in futility.

Revitalization of decadent indigenous cottage industries was given emphasis in the rural reconstruction programme. It could be possible because of the cooperative system. Cash advance and raw materials being the two primary requisites, it was necessary to make both available. Instead of artisans taking loans and advances individually from wholesale traders or moneylenders, Radhakamal advised them to combine and avail of financial support from cooperative credit societies. Weaving, dyeing, tanning, printing and the rest of the handicrafts sector could be modernized. Artisans working in union[65] could install expensive scientific machinery even from abroad. Every region in the country had its distinct brand of product. With proper marketing facilities provided to them, these industries could be given the necessary fillip. This, explained Nagendranath, would also reduce the pressure of population on land even while serving as an alternative source of income. Even the womenfolk could be involved in such enterprises. At their leisure they could be trained to supplement their family income.[66] Practical training in cutting, sewing and embroidery could also be imparted to them. In all such activities, the cooperative principle could serve as the cornerstone.

In fact, in the first quarter of the twentieth century, considerable importance was beginning to be attached to the idea of cooperation. Nityagopal Mukhopadhyay stated that if silk cultivation and business were organized on a cooperative basis he would contribute financially. Silk cultivation required the cooperative effort of a lot of men, good arrangement and expert guidance. Only through *Swadeshi* efforts could the silk industry be improved he said. He even promised to go to Japan and acquire the best cocoons.[67] Charuchandra Das, another rural enthusiast, suggested in 1926 cooperatives among jute farmers.[68] This could check the activities of intermediaries, control price and ensure a fair return for peasants. Radhakamal advocated that scientific laboratories should be set-up in rural areas to enable villagers to be acquainted with modern scientific knowledge, particularly the sort of knowledge related to agriculture, industry and trade, which would provide greater opportunities for financial gains to rural inhabitants. He argued in favour of cooperative societies to experiment with different types of crops and fertilizers in the agricultural farms. Agricultural exhibitions, he suggested, could be organized to enthuse peasants about use of new fertilizers, crops and equipment.[69]

Along with generation of wealth and resources in the countryside equal emphasis was given to the development of the human potential. Radhakamal proposed schools, both day and night, for the peasant children. He wanted literacy for all. To realize this he suggested, books were to be provided free of cost to needy students. There were proposals for organization of recreational and cultural activities, viz., *samkirtans*, *kathakathas*, etc., to promote the community spirit. Nagendranath advocated the need to establish primary schools in each average size village or at least one in close proximity to a couple of neighbouring villages as he was well aware of the reluctance of the simple uneducated rural folk to let their children travel far.

A crucial component in the rural welfare scheme advocated by these young enthusiasts was the health programme. Community health and hygiene was prioritized. It was suggested to divert a part of the village fund to provide drinking water and healthy surroundings. Basic knowledge was needed to be imparted to the common folk to fight common ailments. There were suggestions for primary medical facilities through health centres and dispensaries. Sanitation was to be given due importance to make possible healthy living.

Radhakamal suggested that such a work plan, comprehensive and practical, covering all aspects of rural life, if adopted could add a new dimension to national life.[70] To actualize these programmes, dedicated works were indispensable. Radhakamal had great faith in the educated urban youth and the liberated zamindars and expected them to be active partners in the enterprise. If the villagers could be made self-reliant and placed in a position where they could realize their goals without external help then agriculture and industry, the twin components of the rural economy would be rejuvenated. Employment opportunities in the village would remove the need for urban migration.[71] Once this was attained, Radhakamal was convinced, small factories on cooperative basis could be established under village communities. This would initiate the secondary stage of rural development.[72] The surplus products of these factories could be exported. So an overall improvement in the economic position of the rural population would accompany an increase in productivity.[73]

It thus seems that to these men informed by the intensified nationalist movement and pre-occupation with the issue of nation building, the peasant and agrarian issues were intrinsically linked to the overall rural landscape. There could be no piecemeal solutions to these and the problems faced could not be solved in isolation. There was a growing tendency among them to identify the work of rural reconstruction as an inherent component of the process of nation building as they considered the rural masses as the main constituent of the 'nation'. Solution to rural problems, they held, depended not so much on changes in the land system but on reform of agricultural practices. In this work, they assigned the main role to the rich zamindars and the educated urban youth. However, voices began to be raised against the existing land structure. Dr. Nareshchandra Sengupta, an advocate of Calcutta High Court who later deposed before the Floud Commission, advocated the abolition of the Permanent Settlement.[74] He regarded it as an obstacle to agricultural progress. However, he suggested that the zamindari system be phased out gradually and zamindars be compensated. The eminent scientist Acharya Prafulla Chandra Roy[75] supported him.

In actuality, it seems that in their plans for constructive work they took for granted the tacit support of the government. It is not certain what they proposed to do if the latter reacted in a negative manner. In their proposals for reforms, they argued in favour of nationalist alternatives to colonial institutions. Nevertheless, their programmes were also based on appeals to the good intentions of zamindars and thus rather utopian. It is strange that they should still repose such faith in a class that had shown themselves very unworthy of it in the preceding century. Perhaps the role model of a young enterprising zamindar personified by Rabindranath was always in their subconscious. Neither could they suggest how the distrust of the rural masses towards the educated urban, a fact they were conscious of, could be removed.

It is also significant that in their approach to the peasant question these men did not attempt to consider it as a class that was both stratified and differentiated. The specific problems faced by the different categories were not addressed. The problems of the *bargadars*, which had become so acute by the 1920s that both political leaders and the government were forced to take notice, were generally ignored. While attempting comprehensive plans for rural revival they preferred to treat the peasantry as an undifferentiated whole.

AGRARIAN PROGRAMME AND THE POLITICAL PROCESS

On 3 December 1903, Viceroy Curzon announced in the *Calcutta Gazette*, the government's plan to partition Bengal. During the political movement that followed Rabindranath sought valiantly to draw the attention of the literati towards reform of rural society. But in spite of its varied dimensions, the Boycott and *Swadeshi* movement lacked a peasant programme. It was indifferent to the question of landlord-tenant relationship at a time when the condition of the common peasant was slowly deteriorating. In the memorandum to the government of the Indian Association, signed by Surendranath Bannerjee who had been a vocal protagonist of peasant rights during the marathon rent controversy, the Indian Association requested it to refrain from interference in landlord-tenant relations.[76] The only saving face was the peripatetic poet who symbolized the soul of the struggle. He pleaded for *atmashakti*; necessity of bridging the gulf between the elite and the masses; rural reconstruction; rejuvenation of the village and agrarian revival. But his message was mostly unheeded.

The main support for the movement came from the zamindars. Spread of the movement to *mofussil* areas hindered radical thought on the agrarian question, since the village *bhadraloks* had to depend more on land than the metropolitan intelligentsia. Besides, many of the leaders were often holders of intermediate tenures and thus had a stake in the security of rentals. Therefore, they ignored persistent grievances against high rents and collection of *abwabs*. In fact, Aurobindo Ghosh, one of the brightest luminaries of the period expressly ruled out in his articles on passive resistance the possibility of no-tax campaign as it would alienate the rentier class. This accounted for the apathy of the peasantry towards the movement. Debiprasanna Roychowdhuri, the Brahmo editor of *Nabyabharat*, observed in 1906 that more than nine-tenths of the lower orders were utterly indifferent to *Swadeshi*[77] as they were unable to identify themselves with a movement that expected them to use costlier indigenous products instead of the cheaper foreign variety while refusing to address any of their just demands. However, even he failed to suggest anything more than famine relief and setting up of rural banks to curb moneylending activities.

The leaders were unable to address the vibrant issue of agrarian relationship. The *Bengalee* printed a letter (signed N.) in 1907 detailing the grievances of peasants-arbitrary rent enhancements and the abuses of *bhaoli* or produce rent in Bihar.[78] But in the Bengal Legislative Council, Ashutosh Chaudhuri, a moderate national leader, a leading barrister, a Pabna zamindar, Secretary of Bengal Landholders Association and president of Pabna District Conference[79] criticized the provision for executive intervention in cases of illegal rent enhancement in 1907-8, arguing that 'the tenant can surely get relief from a *munsif*'s court'.[80] Bhupendranath Bose, another moderate nationalist leader, welcomed the summary procedure.[81] Both of them of course, condemned the distinction sought to be drawn between good and bad zamindars. Rabindranath had passionately spoken out in its favour. He himself had set out to model himself on the lines of an ideal zamindar. However, the *Bengalee* was sceptical about the probability of other zamindars following suit.[82] Nor did it view with favour the two petitions from Muslims of Kharaghari village alleging the imposition of an *abwab* of 50 per cent. It was satisfied with the replies of the Muktagaccha zamindars, when the petitions were referred to them by the district magistrate, that 'lately the ryots of this quarter have conceived an idea that their rents will be reduced and that their representation will be favourably considered by the government irrespective of their merits, and probably this idea had also induced them to file the said petition'. The petitions, the zamindars alleged, had failed to pay their legitimate rents regularly, and hence deserved no consideration whatsoever.[83] The apathy to the grievances of the peasantry is all the more glaring in view of rising prices which caused great distress to the ryots.[84]

In this connection, it needs to be mentioned that the landlords helped to extend and enforce the boycott in the countryside as they used their general influence over the ryot tenants or sharecroppers.[85] It was this closeness with the landowning section that was responsible for the inability to take up the peasant cause. Besides, the Bengal Landholders Association played a lead role ably directed by its secretary Ashutosh Chaudhuri. The Indian Association on the contrary failed to organize a volunteer force of its own. This left the ryots without a forum to articulate their grievances.

However, the Association for the Advancement of Scientific and Industrial Education of Indians founded in March 1904 by Jogendrachandra Ghosh planned in August 1905 to set-up an agricultural settlement on 45,000 *bighas* of land purchased near Deoghar. It aimed to provide a centre for training in scientific agriculture. Ultimately, it opened in 1908 with a training school for fifteen sons of gentlemen.[86] In June 1906, an Indian Cotton Cultivation Company was registered to start a cotton farm at Maluti in Santal Parganas to provide raw cotton for the Banga Lakshmi Mill, a *Swadeshi* enterprise. It was an abortive attempt as the crop was ruined by a drought.[87]

These failures boded ill for the future as they strengthened communalist forces. It made it easier for a section of the Muslim aristocratic leadership, to adopt an anti-landlord stance.[88] In Bengal the number of big Muslim zamindars were very few. Nawab Ali Chaudhuri of Mymensingh repeatedly denounced the oppression of Hindu zamindars.[89] *Mihir-O-Sudhakar* that was owned by him suggested establishment of credit societies and revision of interest laws to free peasants from the clutches of moneylenders who constituted the bane of peasant life.[90] It also attacked Hindu rice merchants for buying cheaply from the producers and selling at higher prices.[91] These may be perceived as attempts to locate the sources of exploitation of the Muslim peasantry in particular and alienate them from the Hindu counterparts who were in an equally distressing situation. The popularity of the journal increased thereby having a deep impact on the Muslim mind. On 25 November 1906, a Muhammedan Vigilance Committee was set-up in Calcutta ably guided by Amir Hussain and Sirajul Islam. It intended to establish contacts with district Muslim associations and *anjumans* and thus prevent exploitation of Muslim tenants by their Hindu landlords.[92] The *Bengalee* accused the Vigilance Committee of making 'groundless statements against Hindu landlords'. Yet the growing sympathy of an increasingly powerful section of Muslim society for their co-religionists was undeniable.[93] The Hindu literati lost the chance of portraying themselves as champions of the Muslim masses.

This gave an opportunity for the *Praja* movement in the subsequent period to inform the political aspirations of a substantial section of the Muslim youth. Led by A.K. Fazlul Huq among others, they became the protagonists of the downtrodden Muslim masses.[94] The *Praja* movement took the form of an anti-zamindari agitation and underscored the growing crisis in Bengal's agrarian society. The term *Praja* or tenant came to be commonly applied to all those who had property rights below the zamindars.[95] It included the intermediate tenure-holders, the ryots and the under-ryots, divisions based on the concept of property right.[96] Nevertheless, there were *Praja*s without property rights also, viz., *bargadar*, *adhiar*, *bhag-chasi*, *khetmajur* who were directly involved with production. The leaders were quick to underline the defects of the Bengal Tenancy Act. However, the *Praja* leaders did not present their views on agrarian issues in an organized manner. They presented lists of demands from time to time, outlined their programmes in conferences, meetings, and in the Legislature. Agrarian thinking of their leadership constituted a part of their political programme.

The *Praja* movement mostly swept the eastern Bengal districts where the substantial Muslim peasantry was either underryots or agricultural labourers. It secured the support of a large number of affluent Muslim tenants in the region who wanted to secure land and other means of production to establish their own authority in the rural economy as also in the political sphere. This could only be possible through a struggle against the existing authorities in the rural world, viz., zamindars, moneylenders, traders and upper caste Hindus. They used the communal factor very skilfully. They used, to their advantage, the fact that not only was the major portion of zamindars Hindus, but the officials responsible for estate management too belonged to the same faith.[97] This category of affluent Muslim tenants allied with the newly emerging Muslim professionals, who had close ties with the rural society and mobilized the under-ryots, the *bargadars* and the poor Muslim peasants.

They were soon to channellize the discontent of the Muslims against the zamindari system into a communal conflict. The *Praja* leadership that included some newly emerging professionals fanned some concrete socio-economic grievances. The controversy over the issues of the right to transfer, which was an important incident of a ryot's holding and the ryot's right to trees was a major cause for tenants' resentment. Among other irritants were frequent rent enhancements, excessive *abwabs*, ill-treatment, exploitation of moneylenders and high interest rates.[98] The First World War and its accompanying dislocations worsened the condition of most ryots.

The *Praja* movement started in earnest with the Kamariarchar *Praja* conference in Jamalpur sub-division of Mymensingh district in 1914.[99] Prominent leaders', viz., Fazlul Huq a lawyer from Bakharganj, Maulana Akram Khan from Calcutta and Maniruzzaman Islamabadi, a writer from Chittagong attended it. It forged a close link between local leaders of the *Praja* movement and the Muslim urban professionals. The latter were quick to realize the advantages of securing some form of mass support.[100] The conference circulated the first demand charter of the tenants. It included abolition of zamindars' right to *nazar* and *salami*; reduction of rent; effective measures against illegal exactions by zamindars; occupancy rights to tenants when the land is cultivated by them for twelve years and the landlords should not be entitled to take it away from them because they would not pay higher rent and; tenants' right to plant trees on his land. The charter mainly upheld the rights of better-off ryots and was silent on those of sharecroppers. The *Praja* leaders soon began to form associations to mobilize the Muslim peasants. In December 1917, Fazlul Huq and a group of fellow lawyers and journalists formed the Calcutta Agricultural Association.[101] In early 1920 the Bengal Jotedars and Ryots' Association was founded by leaders from Calcutta and towns of eastern Bengal. At this stage, they demanded abolition of the landlords' fee on transfers of land, abolition of illegal exactions, reduction of rent, the tenants' right to trees, relief from indebtedness and 'honourable treatment of Muslim tenants in the zamindar's *kutcherry*'.[102] Such organized efforts continued through the 1920s. They tried to combine sporadic local actions by peasants against landlords or their agents, or against moneylenders, with the more ordered forms of meeting, processions, boycotts, strikes and refusal to pay illegal cesses. Gradually the *Praja* leaders organized district level associations. These associations did not achieve anything of immediate importance, but they were the precursors of peasant organizations established by Muslim politicians in the mid-1920s, which later provided a backing for Fazlul Huq's Krishak Praja Party in the Legislative Council.

Very soon, Fazlul Huq emerged as the undisputed leader of the *Praja* movement. In 1921, he organized a huge meeting in a village in Barisal. His popularity increased with the success he achieved in the anti-Saha movement in Ghiur in Dacca district in 1926. It was a movement of the poor Muslim cultivators against the local Saha community who were moneylenders-cum-landlords. The former demanded reduction in rent and interest rates.[103] They refused to cultivate the lands of the Sahas and suspended work for the other local Hindu zamindars. The communal factor was becoming a distinct force.

On 3 January 1926 Maulavi Ismail Hussain Siraji, a Muslim jotedar arranged a peasant conference at Salimpur.[104] A fourteen-point charter of demands was formulated here. They included *Kaimi* or permanent right for ryots in their arable and homestead land; unlimited and unqualified right for transfer of ryots' land; zamindars be denied right to dispossess the purchaser thereof; zamindars be denied right to have *khas* possession of land sold after payment of compensation to the purchaser; landlords' fee be realized through registry office with compulsion of zamindars to grant *dakhila* in receipt of same without right to claim *nazar*; provision of increment of rent after lapse of every fifteen years be declared null and void; ryots be given right to valuable trees grown in their lands; ryots be given complete right to excavate tank, sink *pukka* well or erect buildings in their lands without giving any *nazar* and; available lands be always under exclusive possession of the cultivating classes alone. They also demanded that the ryot be able to acquire *Kaimi mourasi* right of his holding if he occupies it continuously for twelve years and the practice of realization of *abwab*s by zamindars be stopped. The zamindars were also to be denied the right to sell the ryots' arable and homestead land, seed store, bullock or the plough for debt.[105] While the Charter reveals the attitude of leaders towards some of the agrarian issues that had become crucial in the politics of 1920s it is evident that the interests of the *jotedars* and affluent peasants were uppermost in their minds. Not much thought was given to the needs of Muslim *bargadars*. Yet the charter received the overwhelming support of Muslim peasantry who viewed the *Praja* movement as the undisputed advocate of their cause. The Hindu peasantry was ignored. The Hindu moneylenders and zamindars were identified as the rural oppressors. By now, there were *Praja samitis* in virtually every district in east and north Bengal. The course of the movement was mainly limited to meetings, processions, suspension of work on lands of Hindu zamindars, refusal of payment of dues, both legal and illegal to zamindars, social non-cooperation with Hindus and sometimes peaceful demonstration before the zamindari *kutcherry*.

What is of considerable significance is the affiliation of the *Praja* leaders, many of whom were of humble origin, in terms of the major political groupings in provincial politics. In contrast with the Western educated Calcutta-based Muslim leadership, they were unequivocally members and supporters of Chittaranjan Das's Congress.[106] But the alliance did not survive his death in 1925 leaving them to feel cheated. In 1926 at the Bengal Provincial Congress Conference[107] the pact was formally annulled.

The Swarajya Party, from its inception in 1923, adopted a pro-zamindar stance. In the election manifesto of 1923, it declared that the agriculturalist of India needed no assurance of the consistent support of the party to his cause. He was the backbone of the country and the mainstay of the Congress. The Swarajya Party would, it observed, miserably fail in its primary duty if it did not make the betterment of his deplorable condition its first and foremost concern.[108] While assuring support to the cause of the agriculturalist, the manifesto added that zamindars in the past had furnished many a brilliant chapter to the country's history and those who desired to help in the building of Swarajya could not possibly dream of undermining the very foundation of the society as it had existed for years by trying to eliminate an important and influential class. However, the Swarajists could not ignore the agrarian issue altogether. When, in the Legislative Council they prepared the work agenda, they decided that instead of waiting for the government to take the initiative in introducing the tenancy bill, they would in consultation with other parties work out a proposal acceptable to both zamindars and peasants and prepare a bill to be introduced in the Council. The leaders of Swarajya Party wanted to keep both agrarian classes satisfied within the framework of the Permanent Settlement. However, after Das's premature death the leaders appointed a committee to realize the plan. It is significant that the social composition of the Swarajya Party would go a long way towards explaining its attitude towards vital agrarian problems. It brought within its fold highly successful Calcutt-based professionals and businessmen, viz., Dr. Pramathanath Banerjee, Saratchandra Bose, C.R. Das, J.M. Sen Gupta and P.D. Himatsingh; big landlords, viz., Harendranath Chaudhuri, Debendralal Khan, Ranjit Pal Chaudhuri and Srischandra Nandy and district organization leaders, usually landlords or professionals and often with terrorist connections, viz., Jogendranath Chakravarti, Jogendranath Moitra and Nagendranath Sen.[109] Besides, in the legislature the Swarajya block often received the support of landlords and businessmen, viz., Sashikanta Acharya Chaudhuri of Mymensingh, Bijoy Prasad Singh Roy of Chakdighi, Bhupendra Narayan Sinha of Nashipur, Satyendrachandra Ghosh Maulik of the British Indian Association and Badridas Goenka of the Bengal Marwari Association. It was because of the social class to which they belonged that the majority of the members of the Swarajya block were defenders of what they termed the 'middle classes of Bengal', i.e. the rentier classes. They favoured the rights of landlords, whatever their position in the hierarchy of proprietorship, against those of under-tenants and sharecroppers.[110] Yet to resolve the differences between the two antagonistic agrarian classes they advocated certain legal changes within the framework of the existing settlement little realizing that failure to find a viable solution to the agricultural problem could lead to more complexities.

The Committee, it was decided, would suggest a solution to the land problem that would be acceptable to both landlords and tenants. The Swarajya Party announced its decision to introduce a Bill in the next session of the Council, based on the proposals of the Committee whose members included Rai Harendranath Choudhury, Rai Satyendranath Choudhury Bahadur, D.N. Roy, Rajibuddin Tarafdar, Kadir Bux and Hemanta Kumar Sarkar. The Committee was asked to submit its report by November 1925.[111] However, the Swarajya Party failed to come to a decision on the land question. Nor could it prepare a report. When in November 1925 it was announced in the *Calcutta Gazette* that a Tenancy Bill would be introduced in the Council, the *Forward*, the mouthpiece of the party, argued in an editorial that the prime reason behind the amendment of the Tenancy Act was to reduce the number of legal suits.[112] It was also suggested that if the right of the zamindars to collect revenue easily were recognized then there would be no objections to recognition of the rights of peasants. But it asserted that there was no provision in the bill which could ensure easy collection of revenue without having recourse to law. Referring to the observation of the committee that relations between zamindars and peasants were cordial, it asserted that the enactment of the Tenancy Bill would adversely affect it. So it concluded that it was unnecessary to initiate changes in existing land system. What it failed to appreciate was that it was the land system that was greatly responsible for the heightening communal tension.

Even the Congress was apprehensive of the changes in the land system that could undermine the existing framework. During the Non-Cooperation movement in 1921-2, a section of the peasantry had joined the struggle. However, class interests prevented the Bengal Congress leaders from providing leadership to the peasants who were justly disillusioned. In 1925, the Bengal Provincial Congress Committee planned to develop the villages by the setting up the Desbhandhu Village Reorganization Fund Committee. But there was no plan to attempt a solution to the land problem even in the context of an impending crisis. The programme adopted by the Committee included establishment of day and night schools, medical relief centres, *charkha* spinning centres, arbitration board for settlement cases, agricultural cooperative and credit societies and cooperative purchase and sale societies.

In the 1924 Congress session at Guwahati, the proposals of the communists included demand for abolition of zamindari system along with Independence for India. But the proposal was defeated by a large majority of votes. The fact that zamindars made significant financial contributions to the Congress coffers could be a possible explanation for this. However, one of the proposals that was accepted was that Congress would try to acquire permanent occupancy right and other such advantages for the peasantry without jeopardizing the interests of the landed classes.

In the meanwhile, the disillusionment suffered at the Krishnanagore Conference gave an opportunity to the Muslim leadership particularly to those who were involved with the *Praja* movement. So long, there had been no attempt to redress the grievances of the Muslim *bargadars*. Considering the latter as a vital force in the conflict with Hindu zamindars and moneylenders, the *Praja* leaders had followed a 'skilful

of the Indian National Congress in December 1927. At the initiative of Muzaffar Ahmad, it laid before the Congress session a detailed programme of abolition of landlordism, of controlling the practice of usury, and of ensuring living wage and democratic rights of combination and strike to the worker.[126] From this time onwards gradually an alliance was forged between them and the left faction of the Congress led by Jawaharlal Nehru. In the third annual conference held at Bhatpara on 31 March and 1 April 1928 it was decided to open a branch of the Peasants and Workers Party at Tangail. The workers there took up the demands of the peasants and spoke of the hardships they had to suffer because of the vast neighbouring area being declared reserved by the government.[127] In this conference, the party's name was changed to the Workers and Peasants Party of Bengal.[128] Ahmad was elected the General Secretary in the executive committee of the party.

The party soon succeeded in organizing peasant protest on justifiable grounds. When the government placed some regions in Atia and Mirjapur in Mymensingh district under the jurisdiction of the forest department without consulting the local cultivators, the Mymensingh branch of the party led by Raijuddin Hussain protested strongly. Muzaffar Ahmad sent a leaflet from Calcutta titled 'An appeal to the cultivators and labourer brethren residing in the Atia Parganas and Tangail' for circulation. He asserted that the lands belonged to the cultivators. The whole incident left Ahmad indignant who continued to support the cause of the affected peasantry by writing on the peasants of Atia and even visiting Dacca and Mymensingh to prove his sympathy for the movement.[129]

In the meantime, an All India Workers and Peasants Party Conference was held in Calcutta from 21-3 December 1928. Here it was resolved that the party could no longer remain content with its existing relationship with the Indian National Congress. It was decided to build-up its own organization. Ahmad moved a resolution proposing the formation of the All India Workers and Peasants Party.[130]

THE INTELLIGENTSIA AND THE EVOLUTION OF THE BENGAL TENANCY (AMENDMENT) ACT, 1928

All this while, the practical working of the Bengal Tenancy Act had revealed serious defects that called for an early remedy. There had been a vast amount of litigation and some conflicting judicial decisions *vis-à-vis* many of the fundamental provisions of the Act. The relationship between zamindars and tenants had deteriorated to the extent as to require legislative interference. In fact, as early as 1900 the British Indian Association observed, 'the land question, in so far as government responsibility to the landlord's interest is concerned, stands thrown back into a comparatively worse situation than what was caused by the legislative changes of 1859'.[131] In a memorandum to the government of Bengal in 1900, the Association listed its grievances against the existing Act—it had practically put a stop to all enhancements of rent and enabled ryots to change the character of holdings on the plea of making improvements.

The Association viewed most seriously the difficulties that landholders were facing even in the matter of realization and recovery by suit in court of admitted rents and cesses. It therefore wanted such amendment of selected clauses in the Act that would afford landholders adequate facilities for the recovery of rents and cesses. However, at that juncture, the government of Bengal was in no mood to initiate any changes in the Act and the Lt.-Governors advised landlords to have faith on judicial decisions.[132]

There were expressions of dissatisfaction with specific clauses of the Act voiced from other quarters too. Section 40 of the Act in particular, under which an occupancy ryot could apply for commutation of his rent-in-kind into money rent, became a constant source of irritation to the settlement officers. They could not accede to the demand to treat *bargadars* as labourers and consequently many of them were recorded as occupancy ryots.[133] The margin between the fixed land revenue and the rent actually payable was the direct cause of not only an increase in tenancy but a number of intermediary interests between the landlords and the actual cultivators. In Bengal, there were instances of as many as fifty or more intermediary interests between the two. The Bengal Tenancy Act provided for occupancy right to be enjoyed by only one person in the chain. The regulations passed in connection with the subsequent amendments to the Permanent Settlement recognized the right in the resident ryots of the village who were generally known as *khud-kasht*. The principle of the settled ryot adopted by the framers of the existing Act had been gradually accepted as a satisfactory recognition of the customary rights of the resident ryot of the village. It afforded a satisfactory solution to the status problem in areas where, conditions were simple and there were only two persons interested in the land, viz., the proprietor landlord and the cultivating tenant. But almost everywhere conditions were rarely so simple. There existed a whole chain of persons interested in the land, both as rent receivers and as rent payers, between the proprietor at the top and the cultivating tenant. Besides, as the law was not properly adapted to the complicated state of sub-infeudation which actually existed, it frequently happened that the occupancy right got into the hands of the wrong person and the cultivating tenant who should have had the right found himself in the position of a tenant-at-will.[134]

The ryots too had for long been asking for a revision of the Act. Khan Bahadur M.A. Momen, magistrate-collector, in his evidence before the Royal Commission on agriculture said in 1928 that 'the ryots' land not being transferable by law, his credit is very small'.[135] Nevertheless, if the land was freely transferable, the cultivator could get more money by mortgaging only 1 *bigha* instead of 3 *bighas*. He was opposed to restrictions on the rights of the ryots to transfer their lands by sale or mortgage.[136]

The clearly visible implications of the Bengal Tenancy Act on rural relationships prompted the government to initiate steps to introduce appropriate legislation. There was a consensus that an amendment to the existing Act was long overdue. However, the war delayed any such attempt. It was only in 1919 that the matter was formally opened in the Bengal Legislative Council. Bishmamadhav Das a member of the Legislative Assembly moved a resolution in 1921 for the appointment of a special committee. The resolution was accepted and a committee was formed with Sir John Kerr as president to suggest amendments. The committee included Raja Bahadur Ban Behari Kapur, Rai Bahadur Surendra Chandra Sen and Sir Ashutosh Choudhury, the well-known High Court judge.

The Report and the Bill of the Committee were published in January 1923 for public debate.[137] The Committee mainly referred to the problems of transferability of occupancy right and the right of the actual cultivator of the soil. The Bengal Tenancy Act had provided that an usage which allowed a ryot to sell his holding without the consent of his landlord would not be affected by the Act. However, Kerr's Committee asserted that this provision was of no use. It was seldom possible for a ryot or his transferee to prove the existence of the usage and there was no guidance given to the courts with regard to the law to be applied to the numerous transfers that were affected without the landlord's consent and without any proof of usage being put forward.[138]

The Committee noted that the number of transfers of occupancy holdings effected by the registered deed had risen and the growing pressure of the population on soil was leading to an ever increasing demand for land. This tended to push up its value and accelerate transfers. Occupancy rights being freely transferred without reference to and without the knowledge of the landlord was a reality. The Committee therefore, recommended that the only remedy was to recognize the practice and to admit the transferability of occupancy holdings subject to the safeguards necessary to protect the interest of the landlords and to secure the general welfare of the agricultural community. The Committee was also in favour of fixing the transfer fee at 25 per cent of the consideration money and to enable the landlord to get a right of pre-emption. The landlord could transfer the holding to himself on payment to the transferor of the consideration money with 10 per cent as compensation, together with any sum which the transferee might have paid in respect of rent or landlord's fee. It proposed to give a limited occupancy right to all under-ryots of whatever grade. For the first time it suggested that, a bonafide cultivator paying a share of the produce to the original owner of the land would be deemed to be a tenant notwithstanding any future contracts to the contrary. For this purpose, a bonafide cultivator was to be defined as a person who himself supplied the plough, cattle and implements of agriculture.[139] The Committee also modified that when the landlord was dependent upon the produce rent for his subsistence, such a rent should not be converted into a money rent. Besides, in view of the disparity that existed between the average value of the rent-in-kind obtained by the landlord and the money rent into which it could be equitably converted some compensation should be payable to the landlord. It also recommended that, the ryot should have complete right in the trees on his holding and proposed that these provisions should be extended to under-ryots with the right of occupancy.

The report and the Bill raised a storm of controversy embroiling the intelligentsia. The general feeling among zamindars and their supporters within the literati was that their class interests were threatened. The British Indian Association resolved to resist any interference with the rights of the landholders.[140] At a conference of landholders held on 3 April at his residence, the *maharaja* of Kasimbazar asserted that the proposed amendment was injurious to the interest of landholders as it aimed to curtail some of their long-standing and just rights in order to benefit the tenants.[141] He felt that, it would lead to deterioration in the relationship between landlords and tenants. The other members present also expressed their misgivings about the possible benefits for peasants if they were given the power of transfer as the chances of the land passing to moneylenders and their being reduced to the position of day labourers from peasant proprietor were great. It was feared that this would enhance ill-feeling between agrarian classes. Therefore, the conference then proposed that the government should either drop the Bill or at least modify some of the crucial clauses.

The pro-landlord press took up the issue and condemned it as being injurious to the interest of all sections—zamindars, ryots and the middle class. The *Barisal Hitoishi*, published since 1896 whose first editor was Rajmohan Chattopadhyay, the former headmaster of Barisal Bangavidyalaya, warned the government of a possible agitation in the country, a 'serious revolution', if the existing Act was amended on the lines recommended. It feared that land owners would replace cultivators by day labourers recruited from outside the province to get their lands cultivated.[142] The *Panchavat* published from Dacca protested against the recommendation to grant occupancy rights in the *barga* lands to *bargadars* and the rights of commutation of the share of the produce into money rent.[143] The weekly *Pratikar* from Berhampore believed that by conferring occupancy rights on sub-tenants the committee was only proposing to reduce the value of the land, which would ensure that tenants were the losers.[144] As a result, wrote the *Mohammadi*, published from Calcutta since 1908 whose proprietor and editor was Maulana Akram Khan, 'litigation would increase beyond control'.[145] The *Navayuga*,[146] also published from Calcutta, and *Desher Vani*[147] reiterated the same opinion. The *Bankura Darpan* and *Jashohar* were more concerned about the effect of the proposed changes on the middle class and the inevitable tension between them and the cultivators, which would result in the pauperization of the latter and the ruin of the former.[148] The *Birbhum Varta* asserted,

> The proposed amendment of the Bengal Tenancy Act, if passed into law, will completely ruin the middle classes. The religions of both Hindus and Muhammedans will be interfered with as this amendment will prejudicially affect the *Devottar*, *Brahmottar* and *Pirottar* lands. This amendment if passed into law will create unprecedented chaos in the villages. Such instances of

robbing Peter and paying Paul have never been found in the history of civilized nations.[149]

The *Swaraj*, published from Calcutta, was even more critical in its attitude. It asserted:

The proposed Bengal Tenancy (Amendment) Bill has got a defect—it leaves some resemblance to the system of the Bolsheviks, civilized society as it stands at present had given the right of ownership to a buyer, and the law of ownership is still being regulated by this principle. Many *jotedars* have bought lands with their money, so if they lost their ownership for letting on *bargas* possibly that will not also be reasonable according to the present definition of ownership. Indeed, the Bill has got no usefulness so far as the tenants or *jotedars* or even society are concerned. If this Bill is passed, it is our conviction that an organized system will be broken, as a result of which many fresh troubles will be created.[150]

Thus, landlords and their protagonists were severely critical of the proposed Bill.

The ryots and their supporters were not far behind in expressing their opinion. In the Bengal Provincial Ryots' Conference, held at Kamarer Char in Mymensingh, a large contingent of ryots and *jotedars* including representatives from the districts were present.[151] Maulavi Khanadaka Naziruddin Ahmed maintained that ryots should be considered as proprietors of the soil and that legislation was imperative to save them from moneylenders. He, however, apprehended that the investing of *bargadars* with occupancy rights and with power to commute produce rent into money rent could lead to unnecessary disputes.[152]

On 23 April 1923 in the concluding meeting of the All Bengal Krishak and Raiyats' Conference held at Calcutta the president Sir P.C. Roy, however, welcomed the Bill.[153] The resolutions passed here included occupancy ryots being given free rights of transfer without paying any *salami* to the superior landlords, the lands of an actual cultivator having a holding consisting of not more than 15 *bighas* of land being made unsaleable in money decree, occupancy ryots having full rights to fell and use trees without paying any *nazar* or *salami* to the land holders, in cases where the tenant himself often effected required improvements, enhancement of rent was not to be allowed on any ground. It was also suggested that where landlords or their agents did not record the actual payment made by a ryot for a particular land, or withheld receipt for which rent was paid, or did not record the full amount tendered by any ryot, or realized *abwab*, fine, etc., his action be counted as a cognizable criminal offence. The Conference also urged the government to issue a circular preventing landlords from filing suits hurriedly for ejecting *jotedars* from the *jots*.

It thus appears that neither zamindars nor protagonists of rights of peasants or the former were in favour of conferring occupancy rights on *bargadars*. As regards the right of transferability, zamindars feared it would cause transfer of land to moneylenders. The British Indian Association, the East Bengal Landholders Association and the North Bengal Zamindars Association favoured modifications in the Bill.[154] On the contrary, others considered the right of transferability as an important concession to ryots.

In view of the violent storm that it had raised, the government considered some changes in the committee's Bill. The main principles were left untouched and were adopted in the Bill that was drawn up by the government in the Revenue Department in 1925. It was introduced in the council on 3 December 1925. It proposed to make occupancy holding transferable subject to payment of certain *salami* or transfer fees to the landlords. Under-ryots would, except in certain cases, be given occupancy rights against their immediate landlord. Tenants were to be given rights to trees, greater facility of payment of rent by money order, the abolition of realization of rent by distraint and the commutation of produce rent into money rent. It was proposed to provide facility for and simplification of the procedure for the realization of rents. The Bill also took into consideration the difficulties of landlords on account of the existence of the co-sharer landlords or co-share tenants.[155]

The changed Bill once again occasioned a prolonged debate in the Legislative Council with zamindars protesting vociferously the pro-ryot clauses. Their main argument was that, if transferability were legalized then ryots' holdings would increasingly fall into the hands of non-agriculturalists. Upholders of interests of ryots were equally dissatisfied. Maulavi Tarafdar, a *Praja* leader, found that 'the proposed Amending Bill will, in reality, make the condition of the tenants worse as the changes are inequitable and in spite of the many advantages enjoyed by, the landlords under the old Act, the amendment will make the landlords enjoy more'.[156] In effect, Maulavi Ekramul Hoque, Shah Syed Emdadul Hoque and other Muslim members in the council adopted a pro-ryot stance. They opposed proposals to give to landlords transfer fee amounting to 25 per cent as well as the right to pre-emption with 10 per cent compensation.

In so far as the question of granting occupancy right to *bargadars* was concerned, all were unanimous in their opposition. Both Ekramul Hoque and Emdadul Hoque argued that special provisions could be made where big landlords were *talukdars* or *jotedars*, letting out land on *barga*, in the law for the purposes of the *barga* system. However, where ordinarily poor but respectable people made a living by letting out their holdings on the *barga* system any provision for accruing rights to the *bargadar* would be extremely mischievous and would foster strife and litigation. They asserted that many lands would remain un-ploughed and those who tilled on *barga* unable to make a living in their villages, would go to the hills, jungles, and perish.[157]

Having failed to satisfy none, the government referred the Bill to a select committee for consideration. It submitted an amended Bill on 22 July 1926. It added a provision to the definition of tenant[158] to clarify that *adhiars*, *bargadars* and *bhag-chasi* were not tenants within the meaning of the Act. It amended the definition of holding, to extend it to undivided shares of parcel or parcels of land. It refused to recommend that under-ryots should be given occupancy rights. However, it suggested that an under-ryot who had held his holding for 20 years and had his homestead was entitled to protection against evictions except on the ground that he had failed to pay the arrears of rent or on the usual grounds on which an occupancy ryot could be ejected. Regarding the right of the ryot to trees, it held that, the landlord should have full right in a limited number of existing trees valuable for their timber which should be specifically mentioned in the Act. All other trees, including all trees that grew or were planted in the future, were to be at the absolute disposal of the ryot. The Committee also modified clause 44 by which a tenant was to be allowed to construct wells, tanks, water channels, etc., for providing drinking water as also erect dwelling houses using masonry bricks, stones or any other materials whatever.

In the select committee, thirteen of the eighteen members sounded a note of dissent. They argued that under-ryots had not been conferred occupancy rights or any substantial rights in lieu thereof. They were convinced that the committee by making holdings of ryots legally transferable and removing from the original Bengal Tenancy Act as well as from the Bill provisions that prevented a co-sharer landlord from becoming a ryot on his own property and by proposing to bar all produce payers, viz., *adhiars*, etc., from obtaining any right as tenants would enable non-agriculturalists to purchase ryoti holdings freely under the assurance that they would never need to cultivate the land themselves. If the actual cultivator delivered part of the produce to them he would be a labourer without any right. If he paid a money rent, he would be an under-ryot with practically no rights under the Act. To avoid such contingencies the dissenters suggested that only if the actual cultivators were given substantial rights in the soil then ryoti holdings could be made legally transferable.[159]

Raja Manmatha Nath Roy of Santosh representing landholding interests observed in his note of dissent,

> The present Bill . . . will virtually lay the foundation of what I may call, "agrarian socialism" and help the so called friends of the ryots to put into their mouth the slogan "land for the tillers alone". Such steps will lead to revolution and not reform. I think it is a short sighted policy to undermine the permanent settlement.[160]

A section of the press raised its voice in their support. The *Pallibasi*, mouthpiece of Burdwan landholders, protested that the proposal to confer occupancy rights on *Korfa* ryots was an 'application of the Bolshevik principle' to the land relations in Bengal: 'This will not only bring about an upheaval in the society and materially ruin the interests of the middle classes but will also decrease the value of the lands by considerably affecting the proprietary rights which have been enjoyed from time immemorial.'[161]

The Muslim intelligentsia however sided with the peasants. The influential *Moslem Jagat* observed that nothing in the Bill was detrimental to the interests of zamindars. In fact, it felt that it was imperative for the tillers of the soil to become the real owners of the land,

> The raiyats want to be free from the clutches of these (the zamindars and the *mahajans*) demons in human shape. They (the raiyats) have taken a solemn pledge that they will not henceforth allow their hard earned money to be spent in the house of prostitutes while their children starve at home. They want permanent rights within lands. Even if the government side with the zamindar the raiyats will not be deterred from their determination for they have now understood that the remedy lies in their hands.[162]

The *Hanafi* wrote, 'It is a matter of wonder that the zamindars of Bengal are in one voice protesting against the Bengal Tenancy

Act (Amendment) Bill, which has given very little privilege to the raiyats while it has considerably enhanced the powers of the zamindars.'[163]

Significantly, protesting against zamindari exploitation the *Praja Bahini* demanded the abolition of the system itself.[164]

In the meanwhile, the government rejected the recommendations of the select committee on the ground that it had radically changed some vital provisions in the Bill. Its main objections to the Bill were that it had no provision for any substantial rights for under-ryots, it declared that no cultivator who did not pay cash rent could be tenant and that, it repealed the provision by which a co-sharer landlord could not hold lands in his own estate on tenure as ryot.[165] The government then decided to refer the matter to a sub-committee consisting of an ex-chief justice of Bengal, Sir Nalini Ranjan Chatterjee and three other officials. The draft Bill of the sub-committee was submitted in July 1927.

The Bill proposed that the full rights of an occupancy ryot should not be given to all under-ryots. Under-ryots who already had the right of occupancy by custom were to be allowed acquisition of occupancy right by law. These under-ryots were to have, as against their immediate landlords, all the rights of occupancy ryots excepting the new right of transferability.[166] The rest of the under-ryots were to be divided into two classes, those who had held their lands, including a homestead, for twenty years continuously as also those who had been admitted in a document by their landlord to have permanent and heritable right and; all others. Both classes were to be liable to ejection on the grounds on which an occupancy ryot was liable to ejection, viz., if they used the land in any way which made it unfit for the purpose of the tenancy or had broken a contract that was consistent with the act and also on one additional ground for arrears of rent. The second class could also be ejected on the ground that their lease had expired on six months' notice. This, however, could only be done if the landlord required the land for his own cultivation. For the purposes of ejection of an under-ryot, cultivation by *bargadar* was not being considered as cultivation by the landlord himself or by hired servants.

The initial rent of the under-ryot was left to contract between the parties. Limits were imposed on subsequent enhancements, viz., if made by contract, not exceeding 4 *anna*s in the rupee and, if by court not to increase the rent to more than one-third of the average gross produce of the previous ten years. Rent once enhanced could not be enhanced again within fifteen years. The Select Committee's suggestion that no class of produce paying cultivators would be treated as tenants prevailed. Only the *dhankararidar*, i.e. persons paying a fixed quantity of produces and also persons already recognized as tenants by their landlords or by the civil courts were excepted. The other *bargadars* or *adhiars* were to be excluded from the definition of tenant. This provision was to apply with retrospective effect even to those cultivators of this class who had been recorded in a record of rights as ryots or under-ryots unless they came under the category mentioned. The provisions in the existing act regarding commutation of produce rent were deleted in accordance with the view of the Select Committee to abolish commutation.

Most of the provisions of the revised Bill were strongly criticized when it was debated in the Legislative Council from August to September 1928. Some members refused to have proposed rights conferred on occupancy rights while others agreed to more restrictive measures.[167] Some of the defenders of zamindari interests fearing curtailment of proprietary rights of landlords if the Bill came into force even demanded that the latter should be compensated otherwise.[168] Surendranath Biswas from Faridpur of the Swarajya Party held that, since the proprietary right was vested in the landlord above they should be in a position to take possession of the land on its transfer by the tenant. Besides, landlords gave out the land to the tenant for enjoyment only and not for transfer. So they should be allowed a share of the value of the land on transfer.[169] Sarat Chandra Basu of the same party reiterated the same fear of agricultural land passing on to banias and marwaris[170] while Ranjit Pal Choudhury was certain, unfettered power if given to ryots would ultimately result in conversion of tenants to *bhag jotedar* or sharecroppers on sale of their holdings.[171]

The pro-ryot members expressed the same objections to the proposal to give 25 per cent of the purchase money by way of *salami* and also of the provision relating to the zamindars' right of pre-emption.[172] It was estimated that since there was transfer of Rs. 2 crore on an average annually, the landlords profited to the extent of Rs. 50 lakh while the transferor could never hope to receive the full value of his property. Thus, landlords, some of whom were not even residents of their zamindaries, would always benefit. The Muslim members in the Council led by Maulavi Tamizuddin Khan of north

Faridpur were unanimous in their fear of greater exploitation of ryots and in their demand for minimum rights for *bargadars.*[173] However, it was over the latter that they were unable to find support from other protagonists of peasant rights who like those of landed interests vehemently opposed any attempt to safeguard sharecroppers and preferred them to be nothing more than wage labourers. Opposition was justified on the ground that generally the owner of the land gave it to the *bargadar* on condition that he would simply cultivate it without any right to it. All liabilities remained with the tenure holder, with the *bargadar* entitled to a share in the produce.[174] Members of the Swarajya Party claimed that to give *bargadars* the right of a tenant was to discourage investment in land.[175] It was also attempted to deprive the labourers of their right of occupancy acquired by them through the settlement records. Attempts by Muslim representatives', viz., Maulavi Nurul Hoque Choudhury of Noakhali and Azizul Hoque of Nadia to reduce the rate of rent of the under-ryot and the *bargadar* were also futile.[176] They argued that considering enhancement of rent could not exceed double the previous rent or one-fourth of the average annual value of the gross produce, produce rent too should necessarily be one-third instead of the one-half. Significantly, all Hindu members including Subhas Chandra Bose voted for the zamindars against the interests of *bargadars*. Rai Sahib Rebatimohan Sarkar, a nominated non-official member, was the sole exception.

THE BENGAL TENANCY (AMENDMENT) ACT

On 14 December 1928, the Bengal Tenancy (Amendment) Act was passed.[177] It was essentially a compromise that belied the latent expectation of the masses. Popular discontent was expressed in contemporary vernacular newspapers and periodicals. The *Ananda Bazar Patrika* warned that the contemplated law in spite of the limited privileges ensured to ryots, had so many provisions in it to put them under the control of zamindars that they could hope for no escape from their miserable plight. The *Saogat* was certain that the new Act would only ensure continual harassment of tenants by zamindars.[178]

After the amendment, it was no longer viable for most Muslim leaders, in their own interest as well as in that of the *praja*, to rely on the Congress.[179] From now on, organized politics among the

Muslim masses of Bengal moved decisively away from the Congress and the leadership of the *Praja* movement decided to form a provincial organization of its own. The Nehru Report and the proceedings of the All Parties Conference in 1928 that drove the final wedge between Jinnah and the Congress accelerated subsequent developments. The all-India political scenario informed Muslim politics in Bengal. The All Bengal Praja Samiti was formed in 1929. In 1931, it resolved to participate in all government institutions, legislatures, municipalities and union boards. This apparently necessitated an appropriate provincial leadership.[180] In the years 1930-4, the Praja Party was controlled at the top by a Calcutta leadership consisting of Sir Abdur Rahim, Khan Bahadur Abdul Momen, Sir Musharraf Hussain and Akram Khan. In 1935 this leadership was challenged when at the Mymensingh conference Fazlul Huq defeated Abdul Momen and became president of the party. From then on the Krishak Praja Party, as it came to be called from 1936, became 'almost entirely an east Bengal party'.[181]

In the 1920s, the organized demands made repeatedly on behalf of the *Praja* movement in eastern Bengal included abolition of illegal exactions reduction of rent and interest rates and relief from indebtedness, honourable treatment of Muslim tenants in the zamindars' estates and abolition of landlords' fee on transfers of raiyati land. The first four demands concerned the peasantry in general. The fifth affected the richer peasantry trying to increase its control over the land and thus demanding a free land market in peasant holdings. Yet, this demand was supported by the majority of peasants, since free transferability and a general rise in land values would have meant easier terms for loans and smaller distress sales of land. In the 1930s, as an impact of the Depression these issues were raised more aggressively leading ultimately to a general demand for the abolition of the Permanent Settlement. Politically, peasant resistance in eastern Bengal did not take the form of an organized *Praja* movement. From 1926-31, there were a series of violent 'communal' clashes against Hindu landlords and moneylenders.

The sense of betrayal flamed by the conduct of the Swarajya Party in 1928 was primarily responsible for the shift in Muslim attitude. Ignoring its election pledge, the former had defended the interest of landlords and sacrificed the just rights of the agriculturists. This is vindicated in the stand taken by its leader Nalini Ranjan Sarkar who in support of landlords' right to *salami* and pre-emption observed,

'I am at a loss to understand, how this can be opposed by those persons who have maintained on the floor of this House that no customary or legal right should be taken away from any body or any person without giving them an adequate compensation.'[182]

The party was overtly anxious to safeguard the interests of the middle class to which many of them belonged. They objected to all attempts to give tenancy rights to *bargadars* on the ground that the middle class would be ruined. The Swarajists also refused to amend provisions relating to the abolition of commutation of produce rent into cash rents, as they considered that it was not just and fair to petty landlords especially when, they were minors, orphans or widows who were incapable of having their lands personally cultivated. It could lead to rising agricultural prices. Nor was it fair in the case of middle class *bhadraloks*, Hindus or Muslims, who could cultivate their lands themselves but let them out for paddy for the consumption of their family just for subsistence, and it was in this light that the matter was considered.[183]

The communists in Bengal too were dissatisfied with the Swarajya Party. Hemanta Kumar Sarkar organized an All Bengal Peasants' Conference at Kustia on 23 and 24 February 1929, to discuss the issue. Muzaffar Ahmad and Nazrul attended it among others.[184] It was decided to organize a meeting of the working committee of the Bengal Peasants League in Calcutta on 24 March 1929, to consider the formation of a branch of the Bengal Peasants League in the districts of Nadia, Khulna, Pabna and Mymensingh. But before the meeting could be held, Muzaffar Ahmad was arrested on 20 March 1929, and sent to Meerut for trial. After this for the next few years following the general party directives the Communist Party in Bengal adopted a different line of action. In the subsequent period they intensified their efforts to win over the poorer sections of the peasantry with some pragmatic and economic demands.[185] Some of these included: the right of pre-emption be immediately withdrawn from zamindars; the transference fee be reduced from 20 per cent to Re. 1 immediately, every person be given occupancy rights to his own land, restrictions be imposed on all enhancements of rent, the present rent be reduced by half, rent be realized not by zamindars but by collectors appointed by the government, the *barga*-holders and agricultural labourers be paid at such rates that their monthly earnings amounted to Rs.40, all rates of interest be limited to 1 per cent and, agricultural banks be established by the government.[186]

NOTES

1. It is a Bengali saying popular in the Birbhum countryside.
2. Abul Hussain, 'Amader Rajniti', *Abul Hussain Rachanbali*, Barna Michil, Dacca, 1976, vol. 1, p. 181.
3. Nazirul Islam Sufian, *Durbipak*, Dacca, 1961, p. 21.
4. Syed Ismail Hussain Siraji, 'Mochalmandiger Daridra Samashyar Samadhan', *Soltan*, Jaisthya 1330 B.S.
5. Ibid.
6. While Nazrul (1898-1976) was acclaimed greatly as a poet he used other literary forms with equal success.
7. *Rekhecho je chal morai bendhe, charti tari pele* (the rice you have stored in the barn, if he gets a handful of it)/*ah-lona mar bhat kheye je bancha esab ehhale*! (He can survive eating it even without salt)/*poshak tomar tarbetarer neiko eder tana* (they have no yearning for good clothes)/ *je-kapare mocho juto, eder tao mele nai* (they don't even get those clothes with which you polish your shoes). Kazi Nazrul Islam, *Nazrul Rachanbali*, ed. Abdul Kader, vol. 1, Kendriya Bangla Unnayan Board, Dacca, 1966, p. 298.
8. *Chashike keu chasha bale* (by calling peasant a peasant)/*kariyo na ghrina* (don't hate him) *banchtam na amra keho* (none of us would have lived)/ *oe se Krishak bina* (without the peasant). Ibid., p. 299.
9. In his '*Samyabadi*' published in 1925.
10. *Bipannader anno thakiye phole mahajan-bhunsi* (depriving the poor of his food the bellies of moneylenders swell)/*nirannader bhite nash kore jamidar chare juri* (by grasping the lands of the hungry, zamindars drive carriages). Kazi Nazrul Islam, op. cit., vol. 2, p. 12.
11. Muhammad Maniruzzaman Islamabadi, 'Anjumane-ulema-O-Samaj Samskar', *Al-Islam*, Asar 1326 B.S., pp. 160-2.
12. Madan appealed to Digambar's son:
    Amar galaye pade damada bar kore fele dan katta/Amar jeno aar eh sayena go, amar jeno aar sayena/Hayre (Allah), ami Kachcha-bachcha, bou-jhi niye konne gedanrabo/ogo apnara daya kare amar hoye babuke duto katha kanpe (Take out the shackles from my feet and neck master/I cannot tolerate anymore/Oh Allah, where will I go and stand with my children, wife and womenfolk/You all please say a few words in my favour to the master). Kazi Imdadul Hoque, *Kazi Imdadul Hoque Rachanbali*, ed. Abdul Kadir, vol. 1, Kendriya Bangla Unnayan Board, Dacca, 1968, p. 36.
    Hoque (1882-1926) did his M.A. in English and became inspector of Mymensingh School and headmaster of Calcutta Training School. He taught in Calcutta and Dacca Madrasa and was the first secretary of board of intermediate and secondary education of Dacca.

13. Abul Hussain, 'Nishader Birambana', op. cit., p. 37.
14. 'Banga Moslemer Durabasthya O Tahar *Pratikar*' (Misery of Bengal Muslims and Its Remedy), *Muazjin*, Baisakh 1335 B.S., p. 16, cited in Shahjahan Munir, *Bangla Sahitye Bangali Musalmaner Chintadhara*, Bangla Academy, Dacca, 1993.
15. *He jamidar, Tumi ki bujhbe se kartabya? Tumi paschimer democracy ke hubohu ene boshiye dio na. Tumi manush hishebe chashar sange boshe jete parbe ki? . . . Tumi chashar klesh bodh korte paro ki? Jodi eshaber konotir shakti tomar na thake, tabe aar democracy-r katha tumi bolo na – vote er janya taka byay kore tomar Praja banglar chashake aar bibrata karona. Tumi tahale bujhlam banglar chasheke balsi kare toolbar jogar karecho . . . Tara manush hoye saman adhikar pete chaye. Tara buker rakta diye utpadan kare sakalke khaoachche kinto nije anshane, ardhashane rin jarjarito hoye udbigno rajani pohachche. Eh aaman abasthya aar katodin?*
    (Oh Zamindar how will you understand that duty? Don't drown us by blindly imposing Western democracy. Can you sit with the ryot like a human being? . . . Can you feel the ryot's suffering? If you don't have the strength for any of these, then don't talk about democracy—by spending money for votes don't disturb the Bengal ryots. Then it will be understood that you are using them as tackles. . . . They want equal rights like human being. They sacrifice their life-blood to feed others but remain hungry themselves, half-starved and in debt they spend nights worriedly. How long will such an unequal situation last?)
    Abul Hussain, 'Banglar Balsi' (Tackles of Bengal), *Bangiya Mussalman Sahitya Patrika*, Sravana 1328 B.S. (1921), pp. 86-7.
16. Abul Hussain, 'Krishi Biplaber Suchana' (Beginning of Agrarian Revolution), ibid, Kartik 1328 B.S. (1921), pp. 224-5.
17. Friedrich List (1789-1846) was a German-American economist. He argued that high protective tariffs were necessary for nations developing new industries. In both the early and advanced stage of a nation's economic development, however, he considered free trade as the appropriate policy. His ideas influenced economic policy in Germany and the US. His best work is *The National System of Political Economy* (1841). *New Standard Encyclopedia*, vol. 8, Standard Educational Corporation, Chicago, 1984, pp. 1-262.
18. Abul Hussain, 'Friedrich List O Tatkalin Germany', *Abul Hussain Rachanavali*, vol. I, op. cit., 1327 B.S., p. 207.
19. Abul Hussain, 'Industrialism: Jantrashilpa Prabaha ba Kaler Karkhana', ibid., 1927, p. 213.
20. Syed Ismail Hussain Siraji, 'Islam O Dharabal', *Al-Islam*, Agrahayan, 1326 B.S., pp. 410-11.

21. Syed Ismail Hussain Siraji, 'Mochalmandiger Daridra Samashyar Samadhan', op. cit.
22. Sadat Ali Akhand, 'Jiban Bima', *Saogat,* Ashwin, 1336 B.S., p. 96, cited in Shahjahan Munir, op. cit., p. 315.
23. Siraji participated actively in the Civil Disobedience movement later.
24. Rehman's (1889-1936) '*Raihan*' was published in 1919. *Raihan*, 3rd edn., Bangla Academy, Dacca, 1978, pp. 82-3.
25. Abul Mansur Ahmed, 'Khaddar Paribo Keno', *Samyabadi*, Falgun 1330 B.S., p. 48, cited in Shahjahan Munir, op. cit.
26. Muhammad Abdur Rashid, *Mussalmaner Artha Sankat O Tahar Pratikar*, Mohammadi Book Agency, Calcutta, 1936, p. 6.
27. *Janagane jara jonksame shoshe tara mahajan raye* (those who suck the blood of people like leeches they are moneylenders)/*santan sama pale jara jani tara jamidar naye* (those who rear peasants like children they are not zamindars)/*matite jader thake na charan* (those whose feet are not on the ground)/*matir malik tanharai hai* (they become the owners of land) in *Sarbahar* published in 1333 B.S.
    Kazi Nazrul Islam, op. cit., vol. 5, 1st half, Bangla Academy, Dacca, 1984, pp. 7-9.
28. *Jai nipirita pran* (Long live the oppressed)/*Jai naba abhijan* (Long live new adventure)/*Jai naba utthan* (Long live awakening), ibid., p. 41.
29. Kazi Nazrul Islam, 'Mandir-O-Masjid', *Ganabani*, 1st year, no. 3, 26 August 1926.
    'Hindu Mussalman', ibid., 1st year, no. 4, 2 September 1920.
30. Sumit Sarkar, *Writing Social History*, Oxford University Press, Delhi, 1997, p. 300.
31. The Bengal Tenancy Act, 1885, did not make many changes in the incidence of occupancy rights. It did not provide for free alienation of holdings. It only imposed certain restrictions on rent enhancements in case of occupancy ryots and the under-tenants of occupancy ryots were left unprotected. Partha Chatterjee, op. cit.
32. Nagendranath Gangopadhyay, 'Palli Samaj Samskar', *Prabasi*, Bhadra, 1324 B.S. (1917), part 21, vol. 1, no. 5, pp. 730-1.
33. Nagendranath Gangopadhyay was the youngest son-in-law of Rabindranath who joined the poet's son Rathindranath at Illinois University in America in 1907 to study the principles and methods of scientific agriculture and animal husbandry. On his return, he joined Tagore's experimental work at Patisar and then Sriniketan. The education he received in America and the first hand experience he acquired enabled him to formulate his ideas on rural reconstruction. His well-informed articles were published regularly in widely circulated *Prabasi* and *Modern Review*.

34. Nagendranath Gangopadhyay, op. cit.
35. Radhakamal Mukhopadhyay was an eminent economist and educationist. After completing his post-graduation, he taught for sometime at Krishnanath College in Behrampore, Murshidabad and then in Calcutta University. He was also Principal of Lucknow University. A prolific writer he was editor of *Upasana*. His works include *Bartaman Bangla Sahitya*, *Monomoy Bharat*, *Taruner Bharat*, *Daridrer Krondon*, and *Visva-Bharat* (2 vols.).
36. Stray efforts, viz., making provisions for food in famine affected regions, occasional medical relief, etc., were often attempted in the rural areas by leaders trying to secure a firm foothold particularly in times of heightened political activities.
37. Bipasha Raha, 'Rural Reconstruction in Early Twentieth Century Bengal: Perception of Nagendranath Gangopadhyay and Radhakamal Mukhopadhyay', *History*, vol. 7, no. 1, 2005.
38. Kali Kumar Mitra, *Prabasi*, Kartik 1335 B.S., part 28, vol. 2, no. 1, p. 51. His personal details are not known. But his writings reveal his deep knowledge about rural society and his interest in agriculture.
39. Ibid.
40. Nagendranath Gangopadhyay, 'Palli Sanskarer Adarsha', *Prabasi*, Bhadra 1325 B.S., part 18, vol. 1, no. 5, pp. 398-400.
41. Bipasha Raha, 'Rural Reconstruction in Early Twentieth Century Bengal: Perceptions of Nagendranath Gangopadhyay and Radhakamal Mukhopadhyay', *History*, vol. VII, no. 1, 2005.
42. Nagendranath Gangopadhyay, 'Palli Samskar Samasya', *Prabasi*, Kartik 1328 B.S., part 21, vol. 2, no. 1, p. 123.
43. In 1904 a credit society was set up. The following figures give an idea of the development of the cooperative movement in the subsequent period:

Year	No. of Cooperative Societies	Members	Capital (Rs.)
(a) 1906	846	91,343	21,31,255
(b) 1911	8,177	4, 03,000	2,02,68,133

Ref. Radhakamal Mukhopadhyay, 'Palli Samskar', *Prabasi*, Bhadra 1320 B.S., part 13, vol. 1, no. 5, pp. 503-4.

44. In every province, a provincial representative committee should be set-up. This committee, covering all the districts, would set-up branches in the villages. Then *mandalis* should be set up, each consisting of a few villages. They would regulate all activities in the villages. This would make self-government a reality. The *mandalis* would initiate establishment of schools, industries, cooperative grain stores and banks in the villages. The heads of the *mandalis* were to be responsible for the welfare of

villages and look after their daily necessities. There should be a meeting place, which would be the centre of all routine activities and recreation in each *mandali*. All disputes within the village were to be settled there through arbitration by the heads of the *mandali*. The *mandalis* would surprise agricultural and artisan production in their respective villages, arrange for the sale of their products at reasonable price as also hoard the grains when the price fell below the fair rate and thus generate prosperity in the villages. Nagendranath Gangopadhyay, 'Palli Samskar Samasya', op. cit.

45. Rabindranath Tagore, 'Pallir Unnati', *Prabasi*, Baisakh, 1322 B.S., part 15, vol. 1, no. 1.
46. Srinath Dutta, a graduate from London, was associated with a number of journals. He was editor of a monthly journal called *Byabshayi*.
47. Srinath Dutta, 'Palligramer Katha', *Prabasi*, Kartik 1325 B.S., part 18, vol. 2, no. 1, p. 75. He outlined here his plan for establishing local self-government. Bengal he wrote had an administrative structure consisting of 27 districts, 84 sub-divisions and 380 *thanas* and a population of 4 crore 54 lakh as to the last census. But this structure was not enough to secure good governance to a rising population.
48. Ibid.
49. Amarnath Dutta, 'Palli Samashya', *Prabasi*, Jaisthya, 1326 B.S., part 19, vol. 1, no. 2, pp. 144-5.
50. Radhakamal Mukhopadhyay, 'Palli Charja Bidhan', *Prabasi*, Magh 1320 B.S., part 13, vol. 2, no. 4, pp. 370-2.
51. He was the Assistant Director of Agricultural Department of Bengal.
52. Nityagopal Mukhopadhyay, 'Shikshita Bhadraloker Krishibritti Abalamban', *Prabasi*, 1309 B.S., part 2, no. 6, pp. 207-8
53. Nityagopal Mukhopadhyay, 'Krishi O Annanya Britti Shiksha', *Prabasi*, 1309 B.S., part 2, no. 1, pp. 6-10.
54. Based on the recommendations of an agricultural meeting held in Shimla, 1903, the Bengal Government in its Regulation I of 1 January 1901 introduced certain provisions for facilitating agricultural education in lower primary, upper primary, middle Bengali and middle English schools. Agricultural science was to be included in the University syllabus. Ibid.
55. After completing his M.A. he joined the Govt. Engineering College at Sibpur. No other personal details are available. But he wrote regularly in the *Prabasi*. His essays reveal his knowledge about agricultural practices and pastoral farming.
56. Dwijadas Dutta, 'Sadharan Krishir Sahit Gopalan O Gabya Byabshayer Tulana', *Prabasi*, Jaisthya 1319 B.S., part 12, vol. 1, no. 2, p. 224.
57. *Prabasi*, 1313 B.S., part 6, no. 1, pp. 646-7.
58. Radhakamal Mukhopadhyay, 'Palli Charja Bidhan', op. cit.

59. It is necessary to refer to the fact that Nagendranath was associated with the cooperative dairy at Sriniketan. He realized the importance of dairy farming for supplementing income of the peasant household. He diagnosed scarcity of green fodder and advocated villagers be acquainted with the techniques of cultivation of seasonal and perennial fodder.
60. Radhakamal Mukhopadhyay, 'Palli Samskar', op. cit.
61. Kali Kumar Mitra, op. cit.
62. Nagendranath narrates an incident that gave him an insight into this. He spent some time in a couple of villages in Bakharganj district during a severe famine. During that time of want and misery, he was witness to the inhuman oppression of local zamindari *kutcherry* officials. They rigorously collected rent from frightened peasants with help of *lathials*.
63. Nagendranath writes that the steep rental demand ensured the perpetuation of a wealthy class who performed no agricultural labour. Most zamindars also preferred to stay in Calcutta or other *mofussil* towns, as they did not find life in the village congenial, thereby leaving estate officials in charge. Absenteeism was responsible for degeneration of villages as all welfare activities were neglected. The rent that was collected mainly sustained ornate urban lifestyles. He believed that *sherista* papers sent to young educated zamindars by their estate officials provided no information about the actual condition of tenants and peasant. This explained their apathy towards conditions in the countryside as they rarely visited their estates.
    Nagendranath Gangopadhyay, 'Zamindar O Krishak Praja', *Prabasi*, Ashwin 1321 B.S., part 14, vol. 1, no. 6, p. 689.
64. Ibid.
65. Radhakamal Mukhopadhyay, 'Palli Charja Bidhan', op. cit.
66. Nagendranath Gangopadhyay, 'Palli Samskar Samashya', op. cit.
67. Nagendranath Som, 'Reshamer Chash', *Prabasi*, 1312 B.S., part 5, no. 10, p. 645.
68. Charuchandra Das, ibid., Falgun 1332 B.S., part 25, vol. 2, no. 5, pp. 632-3.
69. Radhakamal Mukhopadhyay, 'Palli Samskar', op. cit.
70. Radhakamal Mukhopadhyay, 'Palli Charja Bidhan', op. cit.
71. Bipasha Raha, 'Rural Reconstruction in Early Twentieth Century Bengal: Perceptions of Nagendranath Gangopadhyay and Radhakamal Mukhopadhyay', op. cit.
72. Radhakamal Mukhopadhyay, 'Palli Samskar', op. cit.
73. Ibid.
74. *Bharatvarsha*, 1923-4.
75. Ibid.
76. *Bengalee*, 15 February 1907.

77. *Nabyabharat*, Pous 1313 B.S. (December-January 1906).
78. 'The Bengal Raiyats Grievances', *Bengalee*, 23-6 January 1907.
79. 21 June 1907.
80. *Bengalee*, 11 November 1906.
81. Ibid., 16 November 1906.
82. On 13 May 1906, it was reported that Rabindranath and his nephew Surendranath in course of a visit to their zamindari at Shilaidaha had remitted a large part of the rent due to scarcity and arranged loans for the needy. The *Bengalee* raised the question, 'Will Other Zamindars Follow Suit?' *Bengalee*, 30 May 1906.
83. *Bengalee*, 27 and 30 May 1906.
84. Price of grains in Mymensingh rose to Rs.6 from Rs.4 a *maund*. *Bengalee*, 30 May 1906.

    Sumit Sarkar describes the failure to evolve a peasant programme as the 'real Achilles heel' of the entire *Swadeshi* movement. Sumit Sarkar, *The Swadeshi Movement in Bengal, 1903-06*, People's Publishing House, New Delhi, 1973, pp. 333-4.
85. Sumit Sarkar refers to an interesting pamphlet written by Charuchandra Basu Majumdar titled *Bartaman Samashya O Swadeshi Andolan* (Present Day Problems and the *Swadeshi* Movement) published on 1 November 1905. The literary form adopted is that of a dialogue between Ramesh Babu, a *Swadeshi* enthusiast, and Gopi Ghosh, a peasant. The latter at one stage rather timidly suggested that perhaps one way of reducing the poor would be to redistribute the wealth of the rich. The *Swadeshi* leader promptly informs him that no one is really rich in India, rather what everyone should do is to unite and cut down the annual expenditure on foreign goods.

    This says Sarkar is a clear vindication of the reluctance of the leaders to adopt any radical agrarian programme. Ibid., p. 270.
86. *Bengalee*, 30 August 1905, 18 October 1906, 7 March 1908.
87. Ibid., 23 January 1907, 18 April 1904, 8 May 1908.
88. 'The Farce of Hindu-Muslim Union', *Moslem Chronicle*, 26 December 1908.
89. *Mihir-O-Sudhakar*, 15 September 1905, 15 March 1907, 7 February 1908—*RNP* for weeks ending 30 September 1905; 30 March 1907; 15 February 1908.
90. Ibid., 29 September 1905, *RNP* for week ending 7 October 1905.
91. Ibid., 24 August 1906, *RNP* for week ending 1 September 1906.
92. Sarkar, op. cit., p. 444.
93. *Bengalee*, 28 November 1906.
94. Amalendu De, *Pakistan Prastab O Fazlul Huq*, Ratna Prakashan, Calcutta, 1972.
95. Muzaffar Ahmad, *Prabandha Samkalan*, Calcutta, 1970, p. 101.

96. Sachin Sen, *Banglar Raiyat O Zamindar*, Visva-Bharati, Calcutta, 1944, p. 15.
97. The 1921 census report states that the total number of agents, managers of landed estates, clerks, rent collectors, etc., was 1,33,775 out of which 1,10,579 or 82.66 per cent were Hindus and 22,356 or 16.71 per cent were Muslims. W.H. Thompson, *Census of India, Bengal 1921*, vol. V, part II, Calcutta, 1923, p. 362.
98. L.S.S. O'Malley, *Bengal District Gazetteers Rajshahi*, Calcutta, 1916, p. 93.
99. Jatindranath De, 'The History of the Krishak Praja Party in Bengal, 1929-47: A Study of Changes in Class and Inter-Community Relations in the Agrarian Sector of Bengal', unpublished Ph.D. dissertation, University of Delhi, 1971, p. 31.
    This is an extremely informative work on the activities of the Krishak Praja Party. The wide range of sources used by the author has made the work indispensable for the study of the activities of the Krishak Praja Party. I have found it useful in analysis of the representation of peasant grievances at a crucial period of history.
100. J.H. Broomfield, *Elite Conflict in a Plural Society: Twentieth Century Bengal*, University of California Press, Berkeley, 1968, p. 157.
101. *Moslem Hitoishi*, 21 December 1917.
102. Jatindranath De, op. cit., pp. 31-2.
103. *The Mussalman*, Calcutta, vol. XX, no. 20, 16 February 1926, p. 3.
104. Ibid., vol. XX, 7 January 1926, p. 5.
105. H.N. Mitra (ed.), *The Indian Quarterly Registrar*, vol. 1, no. 1, January-March, p. 63, cited in Jatindranath De, op. cit., p. 47, fn. 3.
106. In 1923, the Swarajya Party accepted the Bengal Pact. Representation in the Legislative Council was to be in proportion to the population and through separate electorates. In local bodies, the majority community in each district would have 60 per cent seats and minority community 40 per cent; 55 per cent of government posts to be reserved for the Muslims and; there was to be no music in procession before mosque and cow killing for religious sacrifices was not to be interfered with. However, the Pact was neither wholly supported by the Hindu Swarajists nor by the Muslims as they were not satisfied by the terms.
107. Bengal Provincial Congress Conference was held from 22-3 May 1926 at Krishnanagore. After the annulment of the Bengal Pact, communal riots broke out at Pabna in August 1926 followed by riots elsewhere.
108. *Bengal Legislative Council Progs*, no. 1, 7 August 1928, p. 416.
109. Partha Chatterjee, op. cit., pp. 94-5.

110. Ibid.
111. *Forward*, Wednesday 26 August 1925, p. 3.
112. Editorial headed 'Tenancy Bill', ibid., Tuesday, 24 November 1925, p. 4.
113. Ibid., 22 November 1925; Atulchandra Gupta, *Jamir Malik*, Visva-Bharati, Calcutta, 1351 B.S., pp. 11-12. Sunil Sen, *Agrarian Struggle in Bengal, 1946-47*, People's Publishing House, Delhi, 1972, p. 1.
114. The *Praja Bahini*, the first popular weekly of the *Praja* movement, started by Rajibuddin Tarafdar from Bogra in 1925.
115. Muzaffar Ahmad (1889-1973) a self-educated youth, was associated for long with the Bangiya Mussalman Sahitya Samiti. See Muzaffar Ahmad, *Amar Jiban O Bharater Communist Party: 1920-29*, Dacca, 1972, p. 23.
It was while editing the *Bangiya Sahitya Patrika* that he became acquainted with Nazrul. The two began to publish an evening daily *Navayug*. Ref. Muzaffar Ahmad, *Kazi Nazrul Islam*: *Smritikatha*, Mitra & Ghosh, Calcutta, 1989, p. 3.
There they used to contribute articles on the lives of workers and their problems. Gradually his interest in the labour movement grew. This interest persisted even after he left *Navayug* in 1921, after his dispute with A.K. Fazlul Huq, a lawyer by profession. In early 1920 Ahmad was drawn towards communism inspired by Nalini Gupta, an emissary of M.N. Roy. He was soon arrested with other communists in the Kanpur Bolshevik conspiracy case in 1924 and sentenced to four years imprisonment. However, he was released from jail in early September 1925 on medical ground. On 25 December 1925, he joined the communist conference held in Kanpur where the Communist Party of India was founded. Ahmad was entrusted with the responsibility of organizing the communist movement in Bengal. After his return to Calcutta on 2 January 1926, he became deeply involved in this work.
116. It was founded on 1 November 1925.
117. *Langal*, vol. 1, no. 3, 7 January 1926.
118. G. Adhikari (ed.), *Documents of the History of the Communist Party of India*, People's Publishing House, New Delhi, 1982, vol. II, p. 682.
119. Earlier, in a letter published in *Dhumketu* edited by Nazrul, under the pseudonym of Dwaipayan, Ahmad had expressed his sympathy for the toiling masses. *Dhumketu*, 13 October 1922.
120. *Langal*, vol. 1, no. 3, 14 January 1926. Muzaffar Ahmad, 'The Peasants and Workers Party of Bengal 1927 and 1928', cited in G. Adhikari (ed), op. cit., vol. III.
121. Ibid., vol. IIIA, pp. 155-62.
122. Ibid.
123. Muzaffar Ahmad attempted to familiarize the people with Marxist

ideas. He first used the *Langal* that had to be closed down in the middle of April 1926 because of a fund crunch. After a tireless effort, he succeeded in starting a new weekly *Ganabani* from 12 August 1926. This became the mouthpiece of the Peasants and Workers Party of Bengal. Nazrul and Hemanta Sarkar were regular contributors. From 10 October 1926 the publication of this too had to be stopped due to financial difficulties. It was revived on 14 April 1927.

All this while Muzaffar Ahmad was also busy organizing the Peasants and Workers Party in Bengal.

N.B. Refer Sachinandan Chatterjee, *Muzaffar Ahmad Smriti*, Calcutta, 1988, p. 33.

124. *Ganabani*, 14 April 1927.
125. G. Adhikari, op. cit., vol. IIIB, p. 212.
126. 'Manifesto of the Workers and Peasants Party to the Indian National Congress', ibid., pp. 301-6.
127. Ibid., vol. IIIC, p. 73.
128. Muzaffar Ahmad, Amar Jiban, op. cit., p. 404.
129. Muzaffar Ahmad, 'Peasants of Atia', 'Oppressed Peasants of Atia', 'Atia's Peasants', *Ganabani*, 29 June, 13 July and 31 August 1928 respectively; Muzaffar Ahmad, *Nirbachita Rachana Samkalan*, Calcutta, 1990, pp. 245-7.
130. G. Adhikari (ed.), 'First All India Workers and Peasants Party Conference Reports', op. cit., vol. IIIC, p. 744.
131. Bengal Revenue Department Proceedings, 9 February, P/1-1 of 1900, nos. 31-2, 29 December 1900.
132. Ibid., file no. 585 T-R, 15 May 1902.
133. Kalipada Mitra, *Supplement to the Final Report on the Survey and Settlement in the District of Dacca*, p. 3.
134. Partha Chatterjee, op. cit.
135. Report on the Royal Commission on Agriculture in India, 1928, vol. IV, p. 340, Q22278.
136. Ibid., vol. IV, p. 323.
137. Sir John Kerr Committee's report was published in the *Calcutta Gazette*, 10 January 1923
138. Ibid., p. 4, para 7.
139. Ibid., p. 7.
140. At the Annual General Meeting of the British Indian Association held on 28 March 1923, the President observed:

No one would object to the interest of the agriculturalists and cultivators being duly protected and safeguarded by necessary legislation. But let it not be done by depriving the landlords of the rights and privileges which they, had enjoyed since the days of Lord Cornwallis. The zamindars of Bengal and the neighbouring provinces have always considered the Permanent Settlement of 1793 as their Magna Carta and having stood by the government through good report and evil for nearly a century and a half they expect the government also to stand by them, and to treat their rights as sacrosanct. It would be a very unwise and inexpedient policy to trample upon these rights at this stage and set tenants against their landlords and create a new atmosphere of political unrest in their province. But if the government will persist in its completed legislation, your Association will have to be prepared to put a fight against any interference with the rights of the Bengal landholder. *The Statesman*, 29 March 1923.

141. Ibid., 3 April 1923.
142. *RNP*(B), no. 11 of 1923, p. 223, *Barisal Hitoishi*, 7 March 1923.
143. The *Panchavat* feared that the interest of not only the poor middle class *bhadraloks* but the zamindars and many cultivators would be jeopardized. Ibid., p. 248, *Panchavat*, 9 March 1923.
144. *Pratikar* was of the opinion that the poor alone would suffer if tenancy rights were conferred on holders of *barga* land. Ibid., p. 248, *Pratikar*, 9 March 1923.
145. Ibid., p. 249, *Mohammadi*, 10 March 1923.
146. *Navayug* wrote:
'If the cultivator of *barga* land becomes tenant thereof as it is suggested, many of the land holders will not lease out lands on the *barga* system but import hired labours from the Santal parganas and elsewhere. This will deprive many tenants of their livelihood. Secondly if the right of the transfer is given to the tenants, the zamindar himself can become a purchaser of holdings under him. The result of the bestowal of their right will be that tenants will sell off the lands, which will pass into the hands of capitalists and thus gradually reduce them to the position of labourers'. Ibid., p. 223, *Navayug*, 11 March 1923.
147. Published from Noakhali, ibid., p. 289, *Desher Vani*, 27 March 1923.
148. Ibid., p. 306, *Bankura Darpan*, 1 April 1923 and p. 322, *Jasohar*, 17A.
149. Ibid., p. 355, *Birbhum Varta*, 23 April 1923.
150. Ibid., p. 355, *Swaraj*, 29 April 1923.
151. The conference was held on 29 & 30 March 1923 under the presidentship of Maulavi Syed Nasim Ali.
152. *The Statesman*, 4 April 1923.
153. Ibid., 24 April 1923.
154. Ibid., 3 April, 1923.
155. *Bengal Legislative Council Progs.*, vol. XIV, 3 December 1925, p. 57.
156. Ibid., p. 63.
157. The *Calcutta Gazette*, no. 29 of 1926, part IV, p. 59, 22 July 1926.
158. Clause 13 of Section 3 of the Act.
159. Ibid., pp. 62-9.

160. Ibid., p. 71.
161. *Pallibasi*, Burdwan, 3 February 1926, cited in *RNP*, no. 9, 1926, p. 108.
162. *The Moslem Jagat*, Calcutta, 1 January 1926 cited in *RNP*, no. 2, 1926, p. 26.
163. *Hanafi*, Calcutta, 9 February 1926 cited in *RNP*, no. 10, 1926, p. 125.
164. *Praja Bahini*, Bogra, 25 January 1928, cited in *RNP*, no. 1, 1928 p. 86.
165. The *Calcutta Gazette*, no. 28 of 1928, part IV, p. 94.
166. Clause (d) of Section 160.
167. *Bengal Legislative Council Progs*, no. 1, 7 August 1928, p. 405.
168. Ibid., no. 2, 27 August 1928, p. 677.
169. Ibid., 22 August 1928, p. 437.
170. Ibid., 23 August 1928, p. 505.
171. Ibid., no. 1, 8 August 1928, p. 466.
172. Ibid., no. 2, 2 August 1928, p. 425.
173. Ibid., 30th session (13-18, 20-5, 27-8, 30-1 August), (1, 3-4 September) 1928, vol. XXX, no. 2, p. 19.
174. Ibid., no. 2, 13 August 1928, p. 44.
175. Surendranath Biswas observed that no one would invest money by purchasing land if he realized that by cultivating it on *barga* he would lose his absolute right of enjoying it. Ibid., p. 58.
176. Ibid., 21 August 1928, p. 389.
177. Its chief provisions were:
    (i) The occupancy holdings were declared transferable in whole or part, subject to a transfer fee amounting to 20 per cent of the sale price. The landlord was given a right of pre-emption on payment of the sale price plus 10 per cent as compensation to the purchaser. He also retained the right to levy a fee for the subdivision of holdings in the case of part transfer, because the Act did not make it incumbent on the landlords to divide the holdings in such cases;
    (ii) In order to prevent land from passing to mortgage for indefinite periods, occupancy ryots were allowed to give usufructuary mortgages only for a period of 15 years;
    (iii) Occupancy ryots were given all rights in trees;
    (iv) The right to commute rent-in-kind into cash rent was also abolished;
    (v) Under-ryots were divided into three classes. Under-ryots who had already obtained occupancy rights by custom were given the full rights of of occupancy ryots, except transferability and the right to be deemed protected interests against superior landlord of the ryot. The second class consisted of under-ryots who had

a homestead on their lands and had occupied it for twelve years continuously, or had been admitted in a document by their landlords to have permanent; and heritable right. This class could be ejected if they failed to pay the rent or if they misused the land. The third class of under-ryots could also be ejected on the additional ground that the ryot wanted the land for his own cultivation. The initial rent of under-ryots was left to contract, subject to the provision that it could not exceed one-third of the estimated value of the gross produce. But once their rent had been fixed, it could only be enhanced under a registered contract by 4 *annas* in the rupee;

(vi) Persons who under the system generally known as *adhi*, *barga* or *behag* cultivated the land of another person on condition of delivering a fixed quantity of produce was, however, recognized as tenant, whether he was a ryot or under-ryot as the case might be under the Act.

See Sachin Sen, *Studies in Land Economies of Bengal*, 1935, p. 281.

178. Ibid., p. 391.
179. Abul Mansur Ahmad, *Amar Dekha Rajnitir Panchas Bachar*, Nowroze Kitabistan, Dacca, 1970, p. 61.
180. Jatindranath De, op. cit., p. 92.
181. Ibid.
182. *Bengal Legislative Council Progs.*, 22 August 1928, p. 428.
183. Ibid., 20 August 1928, p. 220.
184. Muzaffar Ahmad, *Kazi Nazrul Smriti Katha*, op. cit., p. 221.
185. Jatindranath De, op. cit., p. 75.
186. *The Sramik*, Calcutta, 28 March 1931, cited in *RNP*, no. 16, 1931, p. 490.
187. *RNP*, no. 29, 1928, 16 August 1928, p. 471.
188. Ibid., 26 August, p. 481.
189. Report on Land Revenue Administration of the Presidency of Bengal, 1929-30, p. 22.
190. Ibid., 1927-8, p. 21.
191. Ibid., 1932-3, pp. 16-17.
192. Ibid., 1937-8, p. 14.
193. Abdul Kasim Fazlul Hoque founded the Nikhil Banga Krishak Praja Samiti in 1927.
194. Amalendu De, *Pakistan O Fazlul Huq*, op. cit., p. 10.
195. *Amrita Bazar Patrika*, 10 August 1936.
196. *Bharatvarsha*, 1933.
197. Ibid.

CHATER 6

Conclusion

The nineteenth-century Bengali literati, which was far from a homogeneous group, were by occupation either salaried or professionals. They preferred professions, viz., law, teaching, journalism and government service to commercial activities. Trade and commerce had little appeal. However, many of them had a share in landownership, as it supplemented income and was also an indicator of social status. A giant portion of their savings was invested in the purchase of landed properties.[1] There was not much change in the composition in the first quarter of the twentieth century. There was greater advancement in Western education and the Western-educated predominated. There was greater preference for liberal professions than for commercial activities. However, the link between land and the literati had weakened considerably. There was greater dependence on education. A considerable section of Muslim society was also able to make its place in the non-homogeneous Bengali literati. Many of them adopted Western education, became qualified professionals and upwardly mobile and were also socially conscious.

As rural society became subject to the fluctuations of the wider capitalist economy and experienced both qualitative and quantitative increases in the economic exchanges in which the rural populace participated, the already awakened social conscious of the literati was bound to take notice. Over the years, the 'peasant' became a subject of study for the litterateurs attempting realistic pen-portraits of rural society they were exposed to either by circumstances of birth or profession. While it all began with Rammohun who reviewed the impact of colonial land policy on agrarian society, literati interest in this domain was sustained over the years by peasant resistance to their subjection, the needs of the anti-imperialist struggle, the changing political scenario, humanistic factors at times, exposure to the realities of peasant life and the impact of world developments, viz., the First World War and the Russian Revolution. The latter rejuvenated interest in the common man in the 1920s.

groups—pro-landlord and pro-tenant. Yet in most cases it is difficult to apply clear-cut categories to a highly educated section. While Harinath was definitely pro-ryot and K.D. Pal, pro-landlord, such categorization does not apply to a Bankimchandra or an R.C. Dutt. Both in their early writings focused on the oppressed peasantry. However, over the years they modified much of their ideas and shifting from their earlier stand expressed overall faith in the potential goodness of landlords.

In this period there was a growing demand for legislative interference to rectify the uncertain nature of peasants' legal position. Increasingly concerned about the growing rural tension there was a general consensus among the literati about the need to convince the government of the urgency for remedial action. Among other issues of interest were the relationship between moneylenders and ryots, the structure of landed society and the general nature of agriculture. It is not that they always offered a coherent solution to all agrarian problems or that they always had an accurate understanding of these problems. However, an attempt was first made to provide organizational support for the ryots. Members of the Indian Association tried to achieve for the ryots what the British Indian Association was trying for zamindars. But the former too was mainly concerned about the ryots with occupancy rights. The 1870s also witnessed a section of traditionally educated Muslim literati attempting to draw the attention of their co-religionists to the peasants' cause. Another significant feature was the reluctance of even peasant sympathizers to support any acts of violence or rebellion against landholders. It was fashionable for the literati with their newly awakened legal sense to hold a public discourse on peasants' rights, rant and rage in the legislature and the press about the rightless condition of a vast multitude of suffering humanity at a time when some of the heinous abuses against a large section of the population—the women—had been mitigated. However, when the peasants themselves rebelled, the world seemed to turn upside down for this class of professionals, most of whom could identify with the landholding classes at a social level but had precious little in common with the actual tillers.

Even the ardent protagonists of ryoti interests fell silent on the issue of 'peasant' and 'agriculture' after 1885. Not even the articulate rural intelligentsia, best represented by Kangal Harinath could carry on a lifelong crusade for the rights of the oppressed peasantry. These were men who lived in the countryside, amongst those for whom they were waging a struggle. As for the rest, Western educated and far removed from the real village, they could never emerge as devouted champions. Their interests were so varied that they could not concentrate on an issue that had held their attention for more than two decades. The Act of 1885 gave them a breathing space and seemed an honourable excuse to shift their priorities. Besides, the shortcomings were not immediately visible. Again, from the 1870s nationalist politics and the need for creating a mass base influenced handling of rural problems.

Yet another significant development of this phase is that literature increasingly became a mode of expression. Much of it read like the romantic pastoral tradition of English literature. Besides, in late nineteenth century here as also in England and France at the same time rural landscape and beauty was gradually related to patriotism:

Gram amader nadir dhare
Aam kanthale ghera,
Teen dike taar mukta math,
Sakal gramer sera!
Aata lebu kul kamala
Bagan bhara kato,
Peyara peach gaachhe gaachhe
Shobhe sata sata!
Aro kata phaler taru
Prati bari bari
Khejur gubak taal narikel
Shobhe sari sari![2]

Rural poverty was also alongside idealizations of the simple way—a theme of fiction in the romantic era and into the twentieth century. Texts composed in this period highlighted this poverty of peasants:

'Nai nai nai' kathar matra eei
Aar katha kichu na shuni kane |
Chhele meyeguli balya khela bhuli
Kande anna tare na sahe pran |
Duti bhaat tare, haath khani dhare
Kandiya kandiya katoi kaye|
'Duti bhaat de ma' 'duti bhaat de ma'
Eh darun katha prane na saye||[3]

A remarkable fact was that there emerged from the late nineteenth century onwards a group of men, unrelated to one another, unsung and unrecognized by the urban literati, away from the callings of nationalist politics, who began to give literary expressions to rural problems, the deteriorating condition of the ryots. They arranged for the publication of these literary creations at their own cost. Many of them were financially solvent with assured income from land or were financed by educated men in their own villages. They were educated, lived in the villages or *mofussil* areas. They were respectable men. Some even had considerable landed interests.[4] They in fact, made spirited attempts to draw the attention of their urban highly educated contemporaries to the sad plight of the peasants:

Mora chashi praja
kandi diba nishi
Amader dukkha kabo kahai |
Amader hoye
duto katha koye
Kare upakar ke achho hai |[5]

What is highly illuminating and significant is there were at times criticisms levelled at the nature of politics practised by the urban literati:

Neta ki 'leader' ki je bole thako
Tao to tomra shunite pai |
Jadi tai hoi tabe keno bolo
Kajer samay khapar pai ?[6]

The latter attempted to portray themselves as representatives of the rural masses. This tone became more pronounced in the post-1885 period particularly, because of the silence on the 'peasant' issue after promulgation of the Tenancy Act. Some of the best known poets', viz., Madhusudan Dutt and Nabinkrishna Sen, steered clear of such divisive issues.

Why was this so? Nabinkrishna was a government official. But then, so were Bankimchandra and R.C. Dutt. It is curious that the lot of the peasantry, the largest section of the population escaped a poet so engrossed with the issues of nationalism and motherland.[7] The late nineteenth century literati used the press extensively to express opinion. However, almost all failed to maintain a non-partisan attitude on the issue of tenant rights. It is also significant that none suggested abolition of the Permanent Settlement and the zamindari system. All reforms that were suggested were within its framework. Besides, the urban literati were not interested in modernizing agriculture. It is curious that while they were acquainted with Western ideas and philosophy, the tremendous improvement and scientific development in agricultural practices achieved in western Europe seemed to have escaped them. Their unfamiliarity with practical agriculture probably accounts for this. While the misery of peasants was visible to them, their lack of basic agricultural knowledge made it difficult for them to understand that the Bengal peasantry still persisted with age-old practices. Securing of tenancy would not solve the basic problem of declining agricultural productivity.

In the period subsequent to the Bengal Tenancy Act, literati interest in such issues subsided. Though it failed to secure the interests of all classes of ryots, all were silenced. Its enactment coincided with the birth of the Congress and political activities received a backseat. It became a deliberate policy of political leaders to stay away from all divisive issues. Besides, a considerable amount of early Congress funds came from the landholding classes. Consequently, no agrarian programme could be adopted. However, enthused by government initiative to promote agricultural education, knowledge and practices, although no comprehensive policy could be adopted at the initial stages, especially after Curzon resumed office a gradually growing number of educated Bengalis became interested. This is evident from the increasing number of agricultural manuals that were published often at private costs.

In early twentieth century, political activities and considerations were uppermost. Political considerations necessitated involvement of the peasantry. This in turn required evidence of concern for their rights. However, closeness of leaders to the landholding classes was perhaps responsible for the failure to evolve a 'peasant' programme even during the Swadeshi and Boycott movement. Yet some little known schemes for agricultural development were undertaken. Rabindranath was the main saving grace. He realized what many of his contemporaries did not, that no real change in the condition of peasants would be possible without overhauling the system of agriculture. Aware of agrarian developments, taking place in the West, he sought to implement them in his estates.

This period saw the beginning of Praja movement. The leaders did not always present their views on agrarian issues in an organized manner. But they tried to combine sporadic action by peasants against landlords or their agents or against moneylenders, with the more ordered forms of meetings, processions, boycotts, strikes and refusal to pay illegal cesses. These associations did not achieve anything of immediate importance but they were the precursors of peasant organizations formed by Muslim politicians in the mid-1920s which later formed a backing for Huq's Krishak Praja Party in the Legislative Council. The Communists were also to use this mass awakening that swept the countryside.

Agrarian thinking became a part of organized politics. The Swarajya Party leaders were sympathetic to zamindars. But Muzaffar Ahmed, the founder of the Communist movement in this region, on the other hand, established the Workers and Peasants' Party to take up the cause of the socially downtrodden. During the amendment of the Tenancy Act controversy, the literati that included an articulate politically motivated group was vocal. The fate of sharecroppers became an issue. But no one was still interested in the masses of agricultural labourers. Their failure to intervene directly in the agrarian class struggle made it impossible for them to emerge as leaders of the peasantry. There was no thought given to an attempt to initiate transformation of the agrarian economy. Ties with the landholding section were still very strong.

A section of the Muslim intelligentsia continued to champion the cause of the peasantry. Their exploitation was consistently discussed in Muslim literature. Hindu litterateurs supporting the ryots' cause often attacked Hindu zamindars. However, the target of Muslim writers was always the Hindu zamindars. The heightening communal tension in the 1920s can be traced to the land question, the disparity in control over land. They kept silent on the fact that condition of Hindu peasants was no better. Nor were Muslim zamindars and officials less oppressive. In general they were not much concerned with the multitudes of Hindu peasants. Yet, simultaneously many of them realized the need for close cooperation with better-off Hindus. There were pleas for cooperative ventures.

The Muslim literati, however, also failed to provide a comprehensive programme for agrarian development. Their agrarian perceptions were primarily limited to highlighting the misery of Muslim peasants. They dealt with issues, viz., agrarian relations, moneylending activities, class antagonism and exploitation of upper classes. But they were silent on issues specifically related to agriculture as an industry, a kind of production system, on which the whole rural structure was based. Vital issues, viz., agrarian yield, tools and techniques of production, scientific farming was overlooked.

This omission is significant at a time when a number of individuals, many highly educated, had become interested in the whole question of agrarian development and rural resuscitation. Many of them had practical knowledge of agriculture. They were aware of the improvements in agrarian practices adopted in Europe and attempted to introduce them here. They undertook experiments and advocated modernization of agricultural practices, to improve productivity, facilitate credit and provide better marketing facilities. They talked about educating the ryots. Without all these, they were convinced, no positive and enduring change in the actual condition of the long suffering peasantry could really happen. Some of the widely read journals, viz., *Prabasi*, *Modern Review* and *Sabujpatra* among others dealt extensively with such issues. Innumerable vernacular tracts on rural themes and agricultural manuals giving crop details were authored, often from the village press. There were instances of the local landed gentry often being involved with such endeavours. One such case was Kalikrishna Choudhury, the zamindar of Bhagirathpur in Murshidabad. He was financially responsible for the publication of a tract titled *Durbhiksha O Daridrata* (Famine and Poverty) authored by Radhikanath Bandyopadhyay.[8] The latter was at one time the editor of *Hindu Ranjika*. This tract first published serially in the paper investigated the reasons for poverty and famine in the country. The book was highly appreciated by lawyers Kaliprasanna Acharya and Prasanna Kumar Bhattacharya and the poet Akshoy Kumar Moitra.

An illuminating feature is that the twentieth-century intelligentsia interested in overall rural and economic development, in their plans for constructive work took for granted the tacit support of the government. It is not certain what they proposed to do in case of the latter's negative response. The exposure to Western influences often facilitated by travels abroad for studies or official purposes, intensification of the nationalist movement, the negotiations the colonial government was often forced to open with political leaders, generated a kind of confidence and inspired thoughts on modernization of the country that involved the masses. However, they were severely handicapped by the fact that even while aware of the mistrust of the peasants towards the educated classes, a feeling often lucidly articulated in verses, who were expected to conduct the constructive work they could not suggest how this mistrust could be removed. Nor could they provide a solution to the basic problem of the land question.

Literati involvement with agrarian issues did not end with

enactment of the Amendment of the Bengal Tenancy Act in 1928. It continued well into the next two decades of colonial rule. Under Fazlul Huq and the Krishak Praja Party agrarian issues were used for mobilization of the Muslim peasantry for political gains. The sharecroppers continued to agitate for some rights. Eminent litterateurs in this period, viz., Tarashankar Bandyopadhyay and Manik Bandyopadhyay among others used agrarian themes for many of their compositions. Tagore's experiments in Sriniketan continued even after his death in 1941. Political mobilization of the peasantry would be complete by the time of the Tebhaga movement initiated in the interest of the sharecroppers on the eve of the transfer of power and Partition of the country. In most literary representation, particularly poetry, the peasant came to be depicted as foolish, ill-fed, ill-clothed and illiterate. The word *chasha* or peasant came to be used in common parlance as one who was uncultured and devoid of social grace.

NOTES

1. Anil Seal, op. cit., pp. 39-42.
2. Our village by the river
   Surrounded by mangoes and jackfruit,
   Open fields on its three sides
   Better than all other villages!
   Custard apples, lemons, berries, oranges
   The orchards are full to the brim
   Guavas, peaches on the trees
   Have grown in hundreds!
   So many other fruit trees
   In every house
   Date palms, beetlenut palms, coconuts
   Are in abundance!
   Kaikobad, 'Amader Gram', *Moslem Bharat*, Year 1, vol. 1, Jaisthya 1327, no. 2.
3. 'Don't have, don't have, don't have' is all that is said
   No other words reach my ears |
   The children forgetting their childhood plays
   Cry that they cannot live without rice |
   For a handful of rice, they hold out their hands
   Crying they keep on saying constantly |
   'Mother give some rice' 'mother give some rice' |
   These terrible words my heart can't bear ||
   Raja Sasisekharesvara Roy, 'Krishaker Prarthana', *Krishaker Chhabi (Abir*

O Anarer Katha), Baikuntha Roy, Tahirpur, 1892, pp. 29-35. See Appendix II for the poem.

4. It is not possible to access the biographical details of all. Raja Sasisekharesvara Roy being a case in point. His *Krishaker Chhabi (Abir O Anarer Katha)*, published from Tahirpur, 1892, is a collection of poems depicting different aspects of peasant life focusing on rural poverty in particular.
5. We ryot tenants cry day-in and day-out
   We shall tell our woes to whom
   On our behalf by saying a few words
   Who is there to help us |
   Raja Sasisekharesvara Roy, 'Sambhranter nikat nibedan', ibid., pp. 23-8.
6. Ibid.
7. Rosinka Chaudhuri, *Gentlemen Poets in Colonial Bengal: Emergent Nationalism and the Orientalist Project*, Seagull Books, Calcutta, 2002.
8. Radhikanath Bandyopadhyay, *Durbhiksha O Daridrata*, Calcutta, 1818 sakabda.

APPENDIX I

Sambhranter Nikat Prarthana

RAJA SASISEKARESVARA ROY

1

Mora chashi praja kandi diba nishi
Amader dukkha kabo kothay!
Amader hoye duto katha koye
Kare upakar ke ache hai |

(We the farmer-subjects do cry day in and day out/Wondering whom to tell of our woes/Who so ever is there to speak for us/Who is there to work for our benefit.)

2

Kar kache jabo kar kache kabo
Kar pa dukhani dharibo kaho
Ja bolo karate sammata tahate
Tabu ekbar edike chaho |

(Whom are we to go and speak of ourselves/Whose feet to touch and request for help/We agree to do whatever you say/Please once focus attention on us.)

3

Mukh tuli chaho amader pane
Murkho dukhi lok jagate mora
Na jani likhite na jani kahite |
Janiye dhalite asrur dhara

(Look at us/The ignorant, sorrow-stricken people of the world/We do not know how to write or speak/All that we can do is to shed tears.)

4

Janire kebol kanditei hai
Kandite kandite janam gelo |
Tobu eto jale mati na bhijilo
Je pashan sei pashan ra'lo |

(We can only shed tears/ Our whole life has been spent in tears/ Yet it was not sufficient to wet the earth/ The rock could not be melted.)

5

Balo balo ohe baro lok sabe
Balo jnanaban bidyan jara |
Tomrao balo shikshita jubak
Bharater lok nahi ki mora?

(Speak out everybody who are rich/ Speak you the wise and literate/ You the educated youth say/ Say if we are not also the citizens of India.)

6

E Bharat bhumi nahe janma sthan
Bharate jeeb mora ki nahi?
Jadi hoi tai tabe keno bhai
'Bhai' bole kabhu dekho na chahi

(Is this country not our motherland/ Are we not creatures of this land/ If that be so then why brother/ Why don't you look upon us as brethrens.)

7

Bhai hote chai ekatha shunei
Hoyeto uthibe ragate jali |
Choto mukhe satya baro katha ei
Kshama karo bhai murkho lok bali ||

(Learning of our claim to be considered your brothers/ You will perhaps be enraged/ It is tall claims from the lowly/ Forgive us for we are ignorant.)

8

Mora murkho lok na jani na bujhi
Bujhi na kichui tobu ore koi
Maramer byatha ke janibe aar
Ke achhe moder tomra boi |

(Illiterate and ignorant we know not/ Neither do we understand or realize/Who else is there to tell our sorrows/We have none but you.)

9

Tomtra bidyan mora murkho lok,
Amader balo 'itar sreni',
Itar sambhrante sambandha kothay?
Tele jale kabhu mele na jani |

(You are educated, we are foolish/ You call us 'lowly classes'/The respectable and the lowly classes can never be amalgamated/Just as water and oil can never mix.)

10

Janiya shuniya tobu hai hai
Prane je bujhe na ta(i) to daki |
Todige na daki dakibo kahai
Keba achhe aar boloto dekhi?

(We know it all/ But our hearts refuse to understand/Who else to ask for help but you/Who else is there.)

11

Amra agyan tomra bidyan
Amra kangal tomra dhani |
Mora bakya hin subakta tomra
Tomra moder rakshak jani

(We are ignorant you educated/We beggars you rich/We inarticulate you orators/We know you are our protectors.)

12

Neta ki 'leader' ki je bole thako
Tao to tomra shunite pai |
Jadi tai hoi tobe keno bolo
Kajer samay khapar pai?

(Leaders or whatever you call yourselves/We come to hear/If that be so then why/Why do we get the news only when we are needed.)

13

Pag bandhi mathe moroli karate
Pratinidhi hoye baktrita dite |
Sabhaye bolite kagaje likhite
Amader bandhu pai shunite ||

(Tying turbans on the head you assert yourselves/As our representatives you give speeches/Speak in meetings write deliberations in newspapers/As our friends.)

14

Jadi tai hoye tabe bolo keno
Na shuno moder dukkher katha?
Jadi bondhu hao tabe keno kao
Na bujho moder praner byatha?

(If that be so then why/Why do you not listen to our woes/If you are a friend then say why/don't you realize our pain.)

15

Anna bina hai asthi charma sar
E dasha moder keno na dekho?
Bharater labane tax diye mori
E Sakal katha kene na likho?

(Foodless we are moving skeletons/Why do you not see our state/We pay taxes our salt/Why do you not write all these.)

16

Malaria jare e desh je galo
Se katha kabhu chinta ki karo?
Pulishe nalishe nashilo je desh
Balo ki karile upay taar |

(We are dying of fever and malaria/Have you ever thought of this/ Police, complaints have battered the country/What have you done.)

17

Keba esakal bujhibe bedona
Maram jatana bujho na hai!
Baje katha loye din kataile
Amra je mori upase kshudai |

(Alas! Who will realize the pain/Who will understand/You are busy with useless issues/While we die of hunger.)

18

Raj durbar hote mushti bhiksha
Paibo e asha peyei hai |
Omni arek sur dhariacho
Amra ekhom kichui noi |

(Alms from the royal court/Assured of receiving it/You have changed your tune/We don't matter anymore.)

19

Mora kichu noi dukkha nahi taye
Dukkha matro ei moder hoye
Baktrita koriye mushti bhiksha peye
Bhojoner kale gale bhuliye |

(We are not sad we don't matter/We are sad that on our behalf/Having made speeches and received a handful of alms/You forget us when eating the morsel.)

20

Mora khudra chasha barbar 'itar'
Kintu nahi tate moder khed!
Bolo dekhi aji ohe shushikshita
Itare sambhrante kato prabhed |

(We are non-entities, farmers, uncivilized/We don't repent/Tell us today you educated people/What is the difference between the respectable you and the uncivilized us.)

(All translation mine.)

APPENDIX II

Krishaker Prarthana

RAJA SASISEKARESVARA ROY

1

'Nai nai nai' katha matra ei
Aar katha kichu na shuni kane |
Chele meyeguli balya khela bhuli
Kande anna tare na sahe pran |

(Not there, not there, not there is all we hear/Our ears hear no other words/The children forgetting childhood games/Cry for food, their wails pierce our hearts.)

2

Duti bhat tare haat khani dhare
Kandiya kandiya katoi koy |
'Duti bhat de ma' 'duti bhat de ma' |
E darun katha prane na saye ||

(They pull their mothers' hand for a morsel/Crying they plead/'Give some food', 'give some food'/The heart cannot bear.)

3

Ki dibo re hai griha shunya moi
Dhaner khosati nahire ghare |
Shak pata chilo taha o phura'lo
Chalar trin o giyache ure ||

(There is nothing in the house/Not even paddy husk/Even the eatable leaves have gone/The roof thatch has blown away.)

4

Sukher sangbad chilore amar
Dhan rakhibar chilo na sthan |
Goale amar balad hajar
Achilo langal eksha khan |

(There was a time when/The barn was too small to hold the grain/Thousand cows filled the shed/More plough shares than I could use.)

5

Chasha bate ami chashitam bhumi
Kintu mor sane parito keba?
Tuchcha rajpad indrer sampad
Rajai korito amar seba |

(A farmer I tilled my land/Who could match me/Government jobs and all the wealth of Indra were insignificant/The king himself served me.)

6

Kintu aji hai! buk fete jai
Dekhiya nijei nijer dasha |
Pete anna nai gaye bastra nai
Sajiachi aaj prakrita chasha |

(But alas today my heart breaks/Seeing myself my condition/No food in the stomach no clothes on the body/Today I am a real farmer.)

7

Onye ashi heshe mor kshetra chashe
Ami hoye dash nirai ghash |
Mor goru hai onye mere khaye-
Coolie hoye bahi tahar mash |

(Others come and till my farm/I weed it as a labourer/While others kill my cows/I carry the flesh as a porter.)

8

Amar pukure onye mach dhare
Ami ghare kore bajare jai |
Bikroy hoile daya kori dile
Jodi ba dukhani ainsh pai |

(Others fish in my pond/I carry the fish to the bazar/If sold the buyer if kind/Gives me some scales.)

9

Ghare ani tai anande dekhai
Chele meye guli tatai khushi |
'mach' 'mach' bole nacher kutuhole-
Cheleder mukhe dhare na hashi |

(I bring it home and happily show/The children are delighted/'Fish' 'fish' they sing and dance with joy/Their happiness knows no bound.)

10

Kintu se hashite amar bukete
Ki bedona je kaha na jai |
Eto chilo jar chele meye tar
Ainshe santushta ajire hai |

(But their joy pierces my heart/I am too hurt to say/One who had so much his children/are satisfied today with only fish scales.)

11

Tobu o e bhalo eto baro bhalo
Ja'hok ta'hok kichuto bate |
Kato din ore ihha o je noi
Iha o adrishte nahiko bate |

(Still it is good it is good/It is at least some thing/Even this was not there for so long/This was also not there.)

12

Saradin jaye kichui na paye
Kandiya kandiya ghumaye pare |

Uthiya abar khunje chari dhar
Pabeto kichure thakile ghare?

(Hungry all day/They cry themselves to sleep/Wake up and again look everywhere/What will they get when there is nothing.)

13

Kichu na paro abasheshe hai
Darun kshudaye garur mato
Ghas pata khaye kintu hai hai
Bujhi mukhe bandhe hoye birata |

(After a futile search/In great hunger like a cow/Eat grass and leaves/It might get stuck in their mouth.)

14

E durdasha aar prane sahe kar
Stree purushe mora nayan mudi |
Punoh khule ankhi chari dike dekhi
Cheleder hoi bipad jadi |

(Our heart grieves at the sight/My wife and I close our eyes/Only to open them and see/The children are not in danger.)

15

Cheleder niye kole basaiye
Ruddha kantha sware kandiya boli |
'Kendona kendona ki achhe bhabona?
Ekhoni bhat ditechi tule ||'

(Taking the children on my lap/Console them in a voice choked with tears/'Don't cry nothing to worry/Give you rice just now'.)

16

Shunya hadi niye unane rakhiye
Mushti bhiksha tare bahir hoi |
Gram gram ghure ashi jabe phire
Jijnase grihini chaul koi?

(Putting the empty rice bowl on the fire/I go out to beg/Moving from village to village I return/My wife queries where is the rice?)

17

Shunya bhiksha thali dui hate tuli
Unmader mato jharite thaki |
Kichu tate na poribe ki chhai
Nahire kshamata aar je likhi

(Holding the empty begging bowl with both hands/Madly I shake it/There is nothing in it/I have no strength to write any more.)

18

Aar ki likhibo? aar ki likhibo?
Aar likhibar kshamata nai |
E hote kathar aro bhayankar
Daridrer chhabi aro ki chai?

(What else to write?/I have no strength to write/There is nothing more horrendous/Do you want more pictures of the indescrible poverty?)

19

Jadi chaho aar bhaire amar |
Daridrer aar o bhishan chhabi |
Krishaker ghare eso daya kore
Dekhibe je chhabi anke na kabi |

(If you want some more brother/Pictures of poverty/Come to a farmer's house/You will see a picture even a poet cannot imagine.)

20

Kabir kalpana je chhabi jane na
Manushe je chhabi bhabite nare |
Heno bhayankar daridra kathor
Dekhibe banger krishak ghare

(A picture a poet can't visualize/One a man can't think of/Poverty more horrendous/You will see in a peasant's hut in Bengal.)

21

Kintu tumi keno tyagi sukh heno
Tyagi ehano swarag such
Sajya sukomal pranayani kol
Ashibe dekhite dukkhir dukkho |

(But why should you sacrifice your happiness/Sacrifice your pleasure/ Leave your soft bed and your beloved/To come to see the misery of the hapless.)

22

Tumiba keno balohe bidyan!
Balo sushikshita! subakta aar
Keno klesh kari griha parihari
Dekhite jaibe dukkha chashar |

(You educated people/Great speakers and orators/Why should you take the trouble to leave home/To see the pain of the farmer.)

23

Sahare thakibe kagaje likhibe
Koribe baktrita bahaba nibe
Chasha mafassale bhashe asru jale
Tahader dukkha bolo keno dekhibe?
—bolo keno dekhibe?

(You will stay in towns, write in papers/Make public deliberations and be applauded/The farmer in the countryside drown in sorrow/Why should you see their pain/Why should you see?)

24

Dekho nare bhai dekhe kaj nai
Tomrai shukhe thakore bhai |
Mora chasha lok kori dukkho bhog
Sukhe thak tora prarthana ei |

(Don't look there is no need to see/You live in happiness/We farmers we live in pain/We pray that you may be happy.)

[Translation mine.]

Glossary of Indian Terms

abwab	traditional arbitrary exaction in addition to formal rent levied by zamindars and other public officers
adhiar	a person sharing half the crop with the landlord
adhyaksha	principal
aguri	middle agricultural Hindu caste
ajlaf	an ordinary Muslim
amla	a petty official
anna	a currency denomination
ashraf	a Muslim of respectable status
ashram	mission
atmashakti	self-reliance
babu	common form of address among the Bengali Hindu gentry; also used pejoratively by British officials in India for the Hindu *bhadralok*s
baidya	a Hindu high caste
bagdi	a low caste Hindu cultivator
Baisakh	a month (mid-April-mid-May) in the Bengali calendar
bandobast	settlement
banik/baniya	merchant
bapari/bepari	contractor
barga	sharecropping
baul	folk singer in Bengal belonging to Vaishnav faith
bauri	a low caste Hindu cultivator
bazar	market
bhadralok	literally 'respectable' but used in historical discourse as an analytical category to imply a status group in Bengal who came from the upper caste; were economically dependent on landed rents and professional and clerical employment and kept a distance from the masses
bhandar	store
bhaoli	produce rent
bhuinya	a lower Hindu caste of cultivators
bigha	a measure of land, ⅓ of an acre

bratidal	group of volunteers
char	alleviated land
charkha	spinning wheel
chaukidar	guard; village police
cutcherry/kutcherry	office of a zamindar
dadan	advance
dalal	broker
daroga	chief policeman
dar patnidar	a subordinate *patni* holder
dar jotedar	a subordinate *jotedar*
dar-us-Islam	holy land for the Muslims
dayabhag	a Hindu law of inheritance prevalent in Bengal
dewan	an official of the zamindar who looks after the finances
dhan	paddy
*dhankararidar*s	persons paying a fixed quantity of produce and also persons already recognized as tenants by their landlords or by the civil courts
dharmaghat	strike
dharmagolas	food granaries
dikus	foreigners
dol	festival of colour
durgotsav	autumn festival
faria	small trader
fatwa	generally written opinion on a point of Islamic law given by theologians or religious leaders
gola	storehouse for grain or salt
gomasta	an agent of the landlord
grihasthya	householder
hat	open air village market usually held twice a week
hool	rebellion
ijaradar	lease-holder
jatra	open air folk theatre in Bengal
jot/jote	a tenure of holding
jotedar	holder of cultivable land; often substantial landholder
jama	rent

kabuliyats/qabuliyat	counterpart to a *pattah*; written agreement given by a tenant to a zamindar assenting to the condition on which he holds land
kangal	beggar
karmisangha	volunteer association of workers
kathakatha	narration of Puranic or other Hindu sacred texts
kayastha	the Hindu writer caste
kayemi/kaimi	perpetual land lease
khal	a creek
khasmahal/khas khamar	personal demesne land
khud kasht	resident tenant
kirtan	Vaishnav musical procession
kist	installment
Koran	Muslim religious text
korfa	tenant-at-will; under-tenant of a ryot
krishibank	agricultural bank
kutcha	opposed to *pukka*; ad hoc; also applied to houses and roads not built of brick or stone
kuthibari	main house; landlord's house
lakh	one hundred thousand
lakhiraj	rent-free land
lathial	armed retainer who mainly used a bamboo stick as a weapon
madrasa/maktab	a higher school or college teaching Islamic laws and jurisprudence as primary subjects
mahajan	moneylender
maharaja	king
maharani	queen
mahishyas	a middle Hindu caste of cultivators
malik	proprietor
mandal	headman
marwari	a person originally from Marwar in Jodhpur state (Rajasthan); marwaris were the richest indigenous mercantile group in Bengal

moulavi/maulavi	a term used for a Muslim doctor of law or a Muslim learned man, also applied to a Muslim gentleman
maund	weight measure 84 lbs.
mirasi/mourasi	hereditary
Mitakshara	Hindu law of inheritance prevalent in north India
mofussil	interior of a district, away from the town or city
mukhtar	lawyers without a formal degree of law but with a license to practice in courts
mukururee/mokarari	fixed; generally referring to a fixed rate of rent
munsif	the lowest grade of judge under British government in India
mutsuddi	broker
namasudra	a low Hindu caste, mostly peasants
naib	a senior official in a zamindar's estate office
nawab	a title or rank conferred like peerage on Muslim gentlemen of distinction and good service
nazar	present/tribute
neij jote	lands cultivated by the proprietors or revenue-payers by themselves and for their own benefit; also land allowed to be set apart for the private maintenance of a zamindar
nirik	standard local rate
pahikasht	non-resident tenant
paik	armed retainer
pukka/pucca	built of brick or stone
panchayat	caste or village council
pargana	a tract of country comprising a number of villages
pathshala	traditional Hindu primary school
patni	a permanent form of management
patnidar	holder of a *patni*
pattah	a document given by a zamindar to a ryot

	specifying the conditions on which land is held
praja	tenant
purohit	priest
raiyat/ryot	peasant, cultivator, tenant
raiyati/ryoti	belonging to a tenant
rajbanshi	a depressed Hindu caste found mostly in the northern districts of the Bengal presidency
reba	interest charged on interest
sabha	association
sadar	headquarter
sadgop	a middle Hindu caste of cultivators
sadhu/sannyasi	ascetic
saha	a title or surname among mercantile social groups in Bengal
sahib	a generic term used to describe an European gentleman
salami	traditional fee to landlord on purchase of land or on obtaining tenancy
samavay	cooperative
samiti	association
sankirtan	collective singing of devotional songs; a Hindu musical procession
sarkar	zamindari official; government
sastra	Hindu religious text
sati	immolation of Hindu widow on the funeral pyre of her husband
satyagraha	a term used to denote the non-violent resistance movement launched by Gandhi against the British
Shariat	Islamic law
sherista	a department or a section of a government office
sheristadar	record keeper in the revenue department
swaraj/swarajya	self-government; political independence
talukdars/talookdar	holder of sub-tenure
teli	Hindu low caste
thana	police station

tol	a school where Hindu scriptures are taught
wahhabi	fundamentalist followers of Abdul Wahhab (1703-87) of Hejiz; the term is applied to fundamentalist agrarian movement in eastern India
waqf	a bequest for religious or charitable purposes, an endorsement, an appropriation of property by will or by gift to the service of god in such a way that it may be beneficial to men, the donor or testator having the power of designating the persons to be so benefited
zamindar	holder of a property in land who paid revenue to the government under the Permanent Settlement of 1793

Bibliography

This work gave me the opportunity to read a wide range of literature. The works of the more well-known authors have been read and re-read and variously interpreted. But there is also a rich mine of vernacular literature produced by lesser known individuals, including tract literature, much of it untapped, often lying in brittle condition in different libraries. They are a great source of information on contemporary social life. I have been able to use some of these rare documents to aquire a glimpse into the manner in which the peasant came to be represented by the contemporary literati.

PRIMARY SOURCES

ARCHIVAL DOCUMENTS/OFFICIAL REPORTS

Act VIII of 1876 [Bengal Legislative Council].

Act VIII of 1885 as modified up to 10 January 1939 (Bengal Government Press, Alipore, Bengal, 1939).

Bengal Act IV of 1928, The Bengal Tenancy (Amendment) Act, 1928, Bengal Secretariat Book Depot, 1929

Bengal Famine Proceedings, 1880.

Bengal Land Revenue (Misc.) Progs., September 1881.

Bengal Land Revenue Proceedings, for the years 1861, 1871, 1876, 1879, 1880-5, 1923-8.

Bengal Legislative Council Progs., for the years 1925-8.

Board of Revenue to the Government of Bengal, Revenue Department, 3 June 1881.

Census of British India of 1871-72, Bengal, London, 1875.

Census of India 1881: Bengal, Calcutta, 1883.

Census of India 1891: Bengal, Calcutta, 1893.

Census of India 1901: Bengal, Calcutta, 1903.

Census of India 1911: vol. 5, *Bengal, Bihar and Orissa and Sikkim*, Calcutta, 1913.

Files of the Home Department (Public, Medical and Sanitary Branches).

Firminger, W.K., *An Introduction to the Bengal Portion of the Fifth Report*, Calcutta, 1917; rpt., Calcutta, 1962.

Garrett, J.H.E., *Bengal District Gazetteer Nadia*, Calcutta, 1910.

General Report on Public Instruction in Bengal (Annual) for 1872-1873 to 1882-1883.

General Report on the Public Instruction in the Lower Provinces of Bengal: Report of the Director of Public Instruction, 1871-2.
Hunter, W.W., *A Statistical Account of Bengal*, London, 1875-7.
Jack, J.C., *Bakharganj Settlement Report (1900-1908)*, Calcutta, 1915.
———, *Faridpur Settlement Report (1904-14)*, Calcutta, 1916.
———, *Economic Life of a Bengal District*, Oxford, 1916.
Land Registration Act, Dacca, 1876.
O'Malley, L.S.S., *Bengal District Gazetteer, Bankura*, Calcutta, 1908.
———, *Bengal District Gazetteer, Birbhum*, Calcutta, 1910.
———, *Bengal District Gazetteer, Hooghly*, Calcutta, 1912.
———, *Bengal District Gazetteer, 24 Parganas*, Calcutta, 1914.
Pal, Kristodas, *Thirty-nine Articles on the Report of the Bengal Rent Law Commission*, Calcutta, 1881.
Peterson, J.C.K., *Bengal District Gazetteer Burdwan*, Calcutta, 1910.
Report of the Land Revenue Commission, 1940.
Report of the Rent Law Commission, 1880.
Report on Land Revenue Administration of the Presidency of Bengal 1927-28, 1929-30.
Speeches of the National Members of the Governor-General's Legislative Council on the Bengal Tenancy Bill 1884, Appendix.
The Extra Supplementary Gazette of India, 1885.
The Gazette of India, 1884.
The Gazette of India, Extraordinary, 1883.
The Imperial Gazetteer of India, vol. XVIII, Indian rpt., New Delhi (Imperial Gazette).
The Supplementary Gazette of India, 1883.
Thompson, W.H., *Census of India, Bengal, 1921*, vol. 5, part II, Calcutta, 1923.

PRIVATE PAPERS

Curzon Papers, India Office Library, Mss. Eur. F. 111/158-230
Richard Temple Papers, India Office Library, Mss. Eur. F. 86

NEWSPAPERS AND PERIODICALS

Relevant issues of:

Amrita Bazar Patrika
Bangadarshan
Bengalee
Bengal Spectator
Bharati

Bharatvarsha
Calcutta Gazette
Calcutta Review
Dhumketu
Ganabani
Hindoo Patriot
Langal
Modern Review
Mussalman
Pioneer
Prabasi
Sabujpatra
Sambad Prabhakar
Santiniketan Patrika
Visva-Bharati Bulletin
Visva-Bharati News

Reports in Native Press (Bengal)

Preserved in the Record Room of the West Bengal Government Secretariat for:

Bankura Darpan
Barisal Bartabaha
Barisal Hitoishi
Bharat Bhritya
Bharat Sanskarak
Birbhum Varta
Biswadut
Dacca Prakash
Dainik Basumati
Desh Hitoishini
Education Gazette
Grambarta Prakashika
Halisahar Patrika
Hanafi
Hindu Hitoishini
Hindu Ranjika
Jasohar
Mihir-O-Sudhakar
Mohammadi
Moslem Chronicle
Moslem Jagat

Nava Yuga
Pallibasi
Panchavat
Praja Bahini
Pratidhwani
Pratikar
Reis and Rayyet
Sadharani
Sahachar
Saptahik Samachar
Som Prakash
Suhrid
Sulabh Samachar
Swaraj

COLLECTION OF NEWSPAPERS

Bandyopadhyay, Brojendranath, *Sangbadpatre Sekaler Katha* (Bengali), vols. I and II, Bangiya Sahitya Parishad, Calcutta, 1377 and 1384 B.S.

———, *Bangla Samajikpatra* (Bengali), vols. I and II, Bangiya Sahitya Parishad, Calcutta, B.S. 1379 and 1384.

Chattopadhyay, Kanailal (ed.), *Samayikpatre Samajchitra, Sanjivani* (Bengali), Dey's Publishing, Calcutta, 1989.

Ghosh, Benoy (ed.), *Selections from English Periodicals*, 7 vols., Papyrus, Calcutta, 1978-81.

———, *Samayikpatre Banglar Samajchitra* (Bengali), 6 vols., Papyrus, Calcutta, 1978-83.

Mamun, Muntassir, *Unish Sataker Bangladesher Sambad Samayikpatra* (Bengali), vols. I-V, Bangla Academy, Dacca, 1985-93.

PROCEEDINGS OF PUBLIC MEETINGS AND ASSOCIATIONS/INSTITUTIONAL PAPERS

All India Congress Committee papers.

Reports and the Proceedings of the British Indian Association available in the library of the Association (relevant issues).

PUBLISHED ORIGINAL LITERATURE/RARE TRACTS

Ahmad, Muzaffar, *Prabandha Samkalan* (Bengali), Mitra & Ghosh, Calcutta, 1970.

———, *Kazi Nazrul Islam: Smritikatha* (Bengali), Mitra & Ghosh, Calcutta, 1989.

———, *Nirbachita Rachana Samkalan* (Bengali), Mitra & Ghosh, Calcutta, 1990.

Al-Hamid Shah, Abdul, *Krishak-vilap* (Bengali), Mymensingh, 1328 B.S.

Al-Samad, Abd, *Krishak Boka*, Abdul Chand Mian, Mymensingh, 1921.

Anonymous, *Adarsha-Krishak* (Bengali), preserved in British Library, Dacca, 1922.

Anonymous, *Falkar*, preserved in the British Library, 4th edn., Calcutta, 1911.

Anonymous, *Sabji-bag* (Bengali), preserved in the British Library, 7th edn., Calcutta, 1917.

Bandyopadhyay, Radhikanath, *Durbhiksha O Daridrata* (Bengali), preserved in the British Library, Gupta Press, Calcutta, 1818 sakabda.

Bandyopadhyay, Harimohana, *Bharatkahini* (Bengali), Kamakhya Charan Bandyopadhyay, Darbhanga, 1900.

Bandyopadhyay, Kedaranatha, *Jeman Deva Temni Devi* (Bengali), preserved in the British Library, Calcutta, 1877.

Bandyopadhyay, Tarasankar, *Rabindranath O Banglar Palli* (Bengali), Mitra & Ghosh, Calcutta, 1971.

Bhattacharya, J.N., *On the Rent Controversy in Bengal*, Calcutta, 1878.

Bose, Nobinkristo, 'The Landed Tenure in Bengal', in the *Proceedings of the Bethune Society for the Session of 1859-1860, 1860-1861*, Bethune Society, Calcutta, 1862.

Chattopadhyay, Bankimchandra, *Bankim Rachanavali* (Bengali), vol. 1, Calcutta, 1360; vol. 2, Calcutta, 1361, Sahitya Samsad edn.

———, 'Samya', English translation—tr. M.K. Haldar in *Renaissance and Reaction in Nineteenth Century Bengal*, Riddhi, Calcutta, 1977.

———, *Sociological Essays: Utilitarianism and Positivism in Bengal*, tr. and ed. S.N. Mukherjee and M. Maddern, Riddhi, Calcutta, 1986.

Chattopadhyay, Sanjibchandra, *Bengal Ryots, their Rights and Liabilities: Being an Elementary Treatise on the Law of Landlord and Tenant* (1st published Calcutta, 1864), rpt., ed. A.C. Banerjee and B.K. Ghosh, K.P. Bagchi & Co., Calcutta, 1977.

———, *Sanjib Rachanavali* (Complete), ed. Asit Kumar Bandyopadhyay, Mondal Book House, Calcutta, 1973.

Chaudhuri, Keshabchandra Acharya, *Rent Question in Bengal*, Calcutta, 1878.

Chaudhuri, Pramatha, *Birbaler Halkhata* (Bengali), Visva-Bharati, Calcutta, 2nd edn., 1926.

———, *Atmakatha* (Bengali), Visva-Bharati, Calcutta, 1946.

———, *Rayater Katha* (Bengali), Visva-Bharati, Calcutta, 1947.

———, *Prabandha Samgraha* (Bengali), vol. 1, 1952, vol. 2, 1954, Calcutta, Visva-Bharati.

———, *Galpasamgraha* (Bengali), Visva-Bharati, Calcutta, 1968.

———, 'Tika', *Sabujpatra*, Asar 1333 B.S.

Das, Abhoycharan, *Indian Ryot, Land Tax, Permanent Settlement and the Famine*, Calcutta, 1881.

Dasa, Anandacharana, *Dhanya Ingrej Raja* (Bengali), Daroka Nath Basu, Barisal, 1877.

Dasa, Kedaranatha, *Bhumyadhikarir Prati Paramarsa* (Bengali), Shyamlal Chakravarty, Azimganj, 1876.

Deuskar, Sakharam Ganesh, *Krishaker Sarva-nasa* (Bengali), rpt. from *Deser katha*, Calcutta, 1908.

Dey, Lal Behari, *Bengal Peasant Life* (1st published in 1874), Macmillan & Co., London; rpt. 1934.

Dey, Prabodhchandra, *Krishikshetra* (Bengali), 2 parts, Calcutta, 1918.

Dutt, Amarnath, 'Palli Samashya', *Prabasi* (Bengali), Jaisthya, 1326 B.S., part 19, vol. 1, no. 2.

Dutt, Dwijadas, Sadharan Krishir Sahit Gopalan O Gabya Byabshayer Tulana', *Prabasi* (Bengali), Jaisthya, 1319, part 12, vol. 1, no. 2.

Dutt, Michael Madhusudan, *Madhusudan Rachanavali* (Bengali), Sahitya Samsad edn., Calcutta, 1965.

Dutt, Romeshchandra, *Bangabijeta* (Bengali), Calcutta, 1874.

———, 'The Bengal Zamindar and Ryot', in *Bengal Magazine*, vol. 2, 1874, under the pseudonym of 'Arcydae'.

———, *On Famines and Land Assessment in India*, B.R. Publishing Corporation, Calcutta, 1985.

———, *England and India: A Record of Progress During the Hundred Years, 1785-1885*, Macmillan, London, 1897.

———, 'The Indian Land Question', *Indian Review*, October, 1902.

———, *Speeches and Papers on Indian Question, 1897-1900*, Elm Press, Calcutta, 1902.

———, *Land Problems in India*, G.A. Natesan & Co., Calcutta, 1902.

———, *Speeches and Papers on Indian Question, 1901-1902*, Elm Press, Calcutta, 1904.

———, *Open Letters to Lord Curzon and Speeches and Papers*, R.P. Mitra, Calcutta, 1904.

———, *Indian Trade, Manufacturers and Finances, etc.*, Elm Press, Calcutta, 1905.

———, *Romesh Rachanavali* (Bengali), ed. Jogeshchandra Bagal, Sahitya Samsad, Calcutta, 1960.

———, *The Economic History of India Under Early British Rule*, Indian rpt., Publications Division, Delhi, 1976.

———, *Economic History of India*, vol. 2: *In the Victorian Age*, Indian rpt., Publications Division, Delhi, 1976.

———, *The Peasantry of Bengal*, ed. Narahari Kaviraj, Manisha Granthalaya, Calcutta, rpt., 1980.

Dutta, Taraknath, *Bhumyadhikari O Prajasambandhiya Ain* (Bengali), Calcutta, 1870.

Gangopadhyay, Nagendranath, 'Palli Samaj Samskar' (Bengali), *Prabasi*, Bhadra 1324, part 21, vol. 1, no. 5.

———, 'Palli Sanskarer Adarsha', *Prabasi*, Bhadra 1325, part 18, vol. 1, no. 5.

Ghataka, Kalimaya, *Lessons in Agriculture*, 2nd edn., Calcutta, 1879.

———, *Krishi-pravesa* (Bengali), Calcutta, 1878.

Ghosh, Ajit Kumar (ed.), *Rammohun Rachanavali* (Bengali), Sadharan Brahmo Samaj, Calcutta, 1973.

Ghosh, Aurobindo, 'Doctrine of Passive Resistance' (Bande Mataram articles, 9-23 April 1907), Sri Aurobindo Ashram, Pondichery, 1948.

Ghosh, Dakshinaranjan, *Krishi-panji O Krishi Samavay* (Bengali), Calcutta, 1916.

Ghosh, Girishchandra, *Girish Rachanavali* (Bengali), ed. Rathindranath Ray and Debipada Bhattacharya, Sahitya Samsad, Calcutta, 1971.

Ghosh, J.C. (ed.), *The English Works of Raja Rammohun Roy*, vols. I and II, Calcutta, 1901.

Ghosh, Manmathanath (ed.), *Selections from the Writing of Girishchandra Ghosh*, the Founder and first editor of *The Hindoo Patriot* and *The Bengalee*, India Daily News Press, Calcutta, 1912.

———, *Raja Dakshina Ranjan Mukhopadhyay*, Manmatha Ghosh, Calcutta, 1917.

Ghosh, Sashisekhar, *Jamidar Darpan* (Bengali), Manager's Office, Narail, 1896.

Gupta, Ishwarchandra, *Kabi Jibani* (Bengali), ed. Bhabatosh Dutt, Sahitya Samsad, Calcutta, 1958

———, *Bhramankari Bandhur Patra* (Bengali), Sahitya Samsad, Calcutta, 1963.

———, *Iswar Gupta Rachanavali* (Bengali), ed. Santi Kumar Dasgupta and Haribandhu Mohanty, Sahitya Samsad, Calcutta, 1974.

Hai, A.F.M. Abdul, *Adarsha Krishak,* Abdul Hai, Mymensingh, 1328 B.S.

Hoque, Kazi Imdadul, *Kazi Imdadul Hoque Rachanavali*, ed. Abdul Kadir, vol. I, Kendriya Bangla Unnayan Board, Dacca, 1968.

Hunter, W.W., *The Indian Mussalmans,* Calcutta, rpt., 1945.

Hussain, Abul, *Abul Hussainer Rachanavali* (Bengali), ed. Abdul Kadir, vol. I, Barna Michil, Dacca, 1383 B.S.

———, 'Banglar Balsi', *Bangiya Mussalman-Sahitya Patrika*, Sravana 1328 B.S.

———, 'Krishi Biplaber Suchana', ibid., Kartik 1328 B.S.

Hussain, Meer Masarraf, *Masarraf Rachana Sambhar* (Bengali), ed. Kazi Abdul Mannan, vol. I, Bangla Academy, Dacca, 1975.

Islam, Kazi Nazrul, *Nazrul Rachanavali* (Bengali), 5 vols., ed. Abdul Kadir, Kendriya Bangla Unnayan Board/Bangla Academy, Dacca, 1966-84 (for some of his works).

———, 'Mandir O Masjid' (Bengali), *Ganabani*, 1st year, no. 3, 26 August 1926.

———, 'Hindu Mussalman' (Bengali) *Ganabani*, 1st year, no. 4, 2 September 1926.

———, *Kavyasamgraha*, ed. Kazi Aniruddha, Kazi Savyasachi and Visvanath De, Mitra & Ghosh, Calcutta, 1973.

———, *Nazrul-vicitra* (Bengali), jt. ed. Kazi Aniruddha, Mitra & Ghosh, Calcutta, 1969.

Islamabadi, Mohammad Maniruzzaman, *Abhibhasan* (Bengali), probably published in 1939

———, 'Anjumane-ulema-O Samaj Samskar', *Al-Islam*, Asar 1326 B.S.

———, 'Samaj Samskar', ibid.

Kaikobad, 'Amader Gram', *Moslem Bharat*, Muhammad Afzal-ul-Hoque, year 1, vol. 1, no. 2, Jaistha 1327.

Mashenullah, M., *Budir Suta* (details of publication not known).

Majumdar, Harinath, *Harinather Granthabali* (Bengali), part I, Jaladhar Sen, Calcutta, 1901.

Mirza, Delawar Hossain Ahmed, *Essays on Mohammedan Social Reform*, vols. I and II, Calcutta, 1889.

Mitra, Dinabandhu, *Dinabandhu Rachanavali* (Bengali), Sahitya Samsad, Calcutta, 1967.

Mitra, Peary Chand, 'The Zamindar and the Ryot', *Calcutta Review*, vol. VI, no. 122, 1846.

———, *Krishi-patha* (Bengali), 3rd edn., Calcutta, 1913.

Mukherjee, Joykrishna (A Lover of Justice), *The Permanent Settlement Imperilled*, Calcutta, 1864.

Mukhopadhyay, Harimohana, *Krishidarpana*, part I, 2nd edn., Calcutta, 1878

Mukhopadhyay, Nilakamala, *Jamidar O Praja* (Bengali), Calcutta, 1873.

Mukhopadhyay, Nityagopal, 'Shikshita Bhadraloker Krishibritti Abalamban', *Prabasi*, 1309, part 2, no. 6.

———, 'Krishi O Annanya Brittishikhsa', *Prabasi*, part 2, no. 1, 1309.

Mukhopadhyay, Prasannachandra, *Palligrama-darpana* (Bengali), Saraswat Press, Calcutta, 1873.

Mukhopadhyay, Radhakamal, 'Palli Samskar' (Bengali), *Prabasi*, Bhadra, part 13, vol. 1, no. 5, 1320.

———, 'Palli Charja Bidhan', *Prabasi* (Bengali), part 13, vol. 2, no. 4, Magh 1320.

Mukhopadhyay, Yadunatha, *Udbhidvichara*, 4th edn., Chikitsa Prakash Press, Chinsurah, 1876.

Naga, Avinasachandra, *Krishak O Sramajibi* (Bengali), Basanta Kumar Mitra, Calcutta, 1907.

Pal, Dwarakanatha, *Udbhidvicharer Prasnottara* (Bengali), Dacca, 1878.

Palit, Haridas, *Gansha* (Bengali), Calcutta, 1915.

Rashid, Muhammad Abdur, *Mussalmander Artha Sankat O Tahar Pratikar* (Bengali), Mohammadi Book Agency, Calcutta, 1936.

Rehman, Muhammad Lutfar, *Raihan* (Bengali), 3rd edn., Ahmed Publishing House, Dacca, 1978.

Rehman, Sheikh Habibur, 'Jamidar O Government' (Bengali), *Islam Darshan*, Agrahayan 1332 B.S.

Roy, Haradhan (ed.), *Vijnan-darpan* (Bengali), vol. I, Calcutta, 1909.

Roy, Parbati Charan, *Rent Question in Bengal*, Calcutta, 1883.

Roy, Raja Rammohun, *Questions and Answers on the Revenue System of India*, London, 19 August 1831.

———, *Paper on the Revenue System of India*, London, 1831.

———, *Remarks on Settlement in India by Europeans*, London, 1832.

———, *Exposition on the Political Operation and the Judicial and Revenue Systems of India*, Calcutta, 1832.

Roy, Raja Sasisekharesvara, *Krishaker Chhabi* (*Abir O Anarer Katha*) (Bengali), Baikuntha Roy, Tahirpur, 1892.

Sarkar, Chandrakanta, *Zamindari Karmadarpana* (Bengali), Calcutta, 1877.

Sarma, Bhagavanchandra, *Krishitattva*, part I, Calcutta, 1869.

Sastri, Haraprasad, *Haraprasad Sastri Rachana Samgraha*, vols. I-IV (Bengali), Paschimbanga Rajya Pustak Parishad, Calcutta, 1989.

Sena, Prasannachandra, *Krishikarjyer Mata* (Bengali), Ishanchandra Seal, Dacca, 1867.

Sen Gupta, Nareshchandra, *Gramer Katha* (Bengali), Calcutta, 1924.

Sen Gupta, Umeshchandra, *Krihichandrika* (Bengali), part I, Tomohur Press, Serampur, 1871.

Sobhan, Sheikh Abdus, *Hindu Mussalman*, Calcutta, 1888.

Siraji, Syed Ismail Hussain, 'Islam O Dhanabal' (Bengali), *Al-Islam*, Agrahayan 1326 B.S.

———, 'Mochalmandiger Daridra Samashyar Samadhan' (Bengali), *Soltan*, Jaisthya 1330 B.S.

Som, Mahendranatha, *Krishak-bala* (Bengali), Dacca, 1914.

Som, Nagendranath, 'Reshamer Chash' (Bengali), *Prabasi*, 1312, part 5, no. 10.

Sufian, Nazirul Islam Mohammad, *Durbipak* (Bengali), Bangla Academy, Dacca, 1961.

Tagore, Rabindranath, *Rabindranath Rachanavali* (Collected Works of Rabindranath Tagore, abbreviated as *RR*), Visva-Bharati Publication, Calcutta, 1971 (for most of his works).

———, 'Europe Prabasir Patra', *RR*, vol. I.

———, *Europe Yatrir Diary*, Bhumika, Calcutta, 1300 B.S.

———, *Chinnapatra*, Calcutta, 1319 B.S.

———, 'Ebar Phirao More', 'Chitra', *Sanchayita*, Calcutta, 1975.

———, 'Prachya O Pratichya', 'Bharatvarsha', *RR*, vol. IV.

———, 'Bharatvarshiya Samaj', 'Atma-shakti', *RR*, vol. III.

———, 'Swadeshi Samaj', 'Atma-shakti', *RR*, vol. III.

———, 'Abasthya O Babostha', 'Atma-shakti', *RR*, vol. III.

———, 'Sabhapatir Abhibhasan', 'Pabna Sanmelani', 'Samuha', *RR*, vol. X.

———, 'Bharatvarsher Itihaser Dhara', *Itihas*, Calcutta, 1362 B.S.

———, *Jivansmriti*, Calcutta, 1319 B.S.

———, 'Pallir Unnati', 'Palliprakriti', *RR*, vol. XXVII.

———, 'Japanyatri', *RR*, vol. XIX.

———, 'Kalantar', *RR*, vol. XXIV.

———, 'Bhumilakshmi', 'Palliprakriti', *RR*, vol. XXVII.

———, *Nationalism*, London, 1918.

———, 'Abhibhasan', 'Palliprakriti', *RR*, vol. XXVII.

———, 'Swaraj Sadhan', Kalantar, *RR*, vol. XXIV.

———, 'Russiar Chithi', *RR*, vol. XX.

———, *The Religion of Man*, London, 1930.

———, Rayater Katha', 'Kalantar', *RR*, vol. XIV.

———, 'Bangalir Kaparer Karkhana O Hater Tant', 'Palliprakriti', *RR*, vol. XXVII.

———, 'Desher Kaj' 'Palliprakriti', *RR*, vol. XXVII.

———, 'Chithipatra', *RR*, vol. IX.

———, 'Sabhyatar Sankat', 'Kalantar', *RR*, vol. XXIV.

———, *The Co-operative Principle*, Compiled by Pulinbehari Sen, Calcutta, 1963.

———, *Galpaguchcha*, vols. I (1397), II (1395), III (1397), IV (1394), Calcutta.

——— 'Samaj', *RR*, vol. XII.

Thakur, Prasannakumar, *Jamidari Karjyer Niyamapatra* (Bengali), Calcutta, 1869.

Vandyopadhyay, Rajendralal (comp.), *A Collection of Agricultural Sayings in Lower Bengal, with Analogous Sayings in Bihar and Orissa*, Calcutta, 1893.

Vidyasagar, Iswar Chandra, *Vidyasagar Rachana Samgraha* (Bengali), ed. Gopal Haldar, Mitra & Ghosh, Calcutta, 1972.

Watt, George, *First Step in Botany: Being an Introductory Treatise on the Vegetation of Bengal*, tr. Dwarakanatha Chakravarti, *Udbhidvidyar Prathama Sopana*, Calcutta, 1876.

AUTOBIOGRAPHY, BIOGRAPHY, MEMOIRS AND SPEECHES

Ahmad, Muzaffar, *Amar Jiban-O-Bharater Communist Party, 1920-1929* (Bengali), Bangla Academy, Dacca, 1972.

Banerjee, A., *Speeches by Lalmohan Ghosh*, parts I and II, Calcutta, 1883 and 1884.

Banerjee, S.N., *A Nation in Making: Being the Reminiscences of Fifty Years of Public Life*, Oxford University Press, Bombay, 1966.

Basu, A.N., *Mahatma Sisir Kumar Ghosh* (Bengali), Calcutta, 1327 B.S.

Bose, K.C., *A Lecture on the Life of Ramgopal Ghosh*, K.C. Bose, Calcutta, 1868.

Chatterjee, Sachinanda, *Muzaffar Ahmad Smriti* (Bengali), Mitra & Ghosh, Calcutta, 1988.

Collette, Sophia Dobson, *The Life and Letters of Raja Rammohun Roy*, ed. Dilip Biswas and Prabhat Chandra Ganguli, Sadharan Brahmo Samaj, Calcutta, 1962.

Hussain, Mir Masarraf, *Amar Jibani* (Bengali), ed. Debipada Bhattacharya, Mitra & Ghosh, Calcutta, 1384 B.S.

Kripalini, Krishna, *Rabindranath Tagore: A Biography*, 2nd edn., Visva-Bharati, Calcutta, 1980.

Mitra, C.C., *Speeches of Ramgopal Ghosh*, Calcutta.

Mitra, K.C., *Memoirs of Dwarkanath Tagore*, K.C. Mitra, Calcutta, 1870.

Pal, Bipan Chandra, *Memories of My Life and Times*, Modern Book Agency, Calcutta, rpt., 1932.

Palit, R.C. (ed.), *Speeches of Surendranath Banerjee 1876-1880*, Calcutta, 1880.

Sanyal, Ramgopal, *A General Biography of Bengal Celebrities: Both Living and Dead*, Calcutta, 1889.

———, *The Life of the Hon'ble Rai Kristo Das Pal Bahadur*, Calcutta, 1886.

———, *Harish Chandra Mukhopadhyay Jibani* (Bengali), Calcutta, Ramgopal Sanyal, 1887.

Sarkar, H.C., *Life and Letters of Raja Rammohun Roy*, Calcutta, 1911.

Sastri, Sibnath, *Atmacharit* (Bengali), Sadharan Brahmo Samaj, Calcutta, 1991.

Tagore Debendranath, *Atmajibani* (Bengali), Visva-Bharati, Calcutta, 1962.

Tagore, Rathindranath, *Pitrismriti* (Reminiscences of My Father), Visva-Bharati, Calcutta, 1968.

———, *On the Edges of Time*, Visva-Bharati, Calcutta, 1981.

———, *Father as I Knew Him: Centenary Volume, 1861-1961*, Visva-Bharati, Calcutta, 1961.

PUBLISHED SECONDARY WORKS/UNPUBLISHED THESES

Adhikari, G. (ed.), *Documents of the History of the Communist Party of India*, People's Publishing House, Delhi, 1982.

Adhikari, Sachindranath, *Shilaidaha O Rabindranath* (Bengali), Jijnasa, Calcutta, 1974.

———, *Rabindramanaser Utsa Sandhane* (Bengali), Jijnasa, Calcutta, 1976.

Adhikari, Santosh Kumar, *Vidyasagar: Education, Reformer and Humanist* (Bengali), Sahityika, Calcutta, 1995.

Ahmed, Abul Mansur, *Amar Dekha Rajnitir Panchas Bachhar* (Bengali), Nowroze Kitabistan, Dacca, 1970.

Ahmed, Aijaz, *In Theory: Classes, Nations, Literatures*, Verso, London, 1992.

Ahmed, Rafiuddin, *The Bengal Muslims, 1841-1906: A Quest for Identity*, Oxford University Press, New Delhi, 1981.

Ahmed, A.F. Salauddin, *Social Ideas and Social Change in Bengal: 1818-35*, Riddhi, Calcutta, 1976.

Ahmed, Sufia, *Muslim Community in Bengal, 1884-1912*, Oxford University Press, Dacca, 1964.

Alavi, Hamza and Theodar Shanin (eds.), *Introduction to the Sociology of Developing Societies*, Macmillan, London, 1982.

Alphonso-Karkala, John B., *Indo-English Literature in the Nineteenth Century*, University of Mysore Press, Mysore, 1970.

Anisuzzaman, *Muslim-Manas O Bangla Sahitya* (Bengali), Lekhak Sangha Prakashani, Dacca, 1968.

Ascoli, F.D., *A Revenue History of the Sunderbans from 1870-1920*, The Bengal Secretariat Book Depot, Calcutta, 1921.

Ashcroft, Bill et al., *The Empire Writes Back: Theory and Practice in Post-Colonial Literatures*, Routledge, London, 1989.

Baden-Powell, B.H., *The Land Systems of British India* , vols. I-III, Oxford University Press and Steven & Sons Ltd., London, 1892.

Bagal, Jogesh Chandra, *History of the Indian Association 1876-1951*, Bharati Library Publishers, Calcutta, 1953.

———, *Peasant Revolution in Bengal*, Bharati Library, Calcutta, 1953.

———, *Unobingsha Satabdir Bangla* (Bengali), Bharati Library, Calcutta, 1963.

———, *Muktir Sandhane Bharat: Congress Purva Yuga* (Bengali), Bharati Library, Calcutta, 1972.

Bagchi, Amiya Kumar, *Private Investment in India, 1900-1939*, Cambridge University Press, Cambridge, 1972.

Banerjee, Anil Chandra, *The Agrarian Systems of Bengal*, vol. 2, K.P. Bagchi, Calcutta, 1981.

Bandyopadhyay, Arun and B.B. Chaudhuri (eds.), *Tribes, Forest and Social Formation in Indian History*, Manohar, Delhi, 2004.

Bandyopadhyay, Brojendranath, *Kalikata Sanskrita Colleger Itihas* (Bengali Publication Marking the 125th Anniversary of the College), part I; *1824–1858*, Sanskrit College, Calcutta, 1948.

———, *Sahitya Sadhak Charitmala* (Bengali), enlarged 4th edn., Bangiya Sahitya Parishad, Calcutta, 1353 B.S.

Bandyopadhyay, Chandicharan, *Vidyasagar* (Bengali), Sahitya Samsad, Calcutta, 1302 B.S.

Bandyopadhyay, Saroj, *Bangala Upanyaser Kalantar* (Bengali), Calcutta, 1988.

Bandyopadhyay, Sekhar (ed.), *Rethinking Bengal History: Esasys in Historiography*, Manohar, Delhi, 2001.

Banerjee, Sumanta, *The Parlour and the Streets: Elite and Popular Culture in Nineteenth Century Calcutta*, Seagull Books, Calcutta, 1989.

Banerjee, Swapna M., *Men Women and Domestic: Articulating Middle-Class Identity in Colonial Bengal*, Oxford University Press, Delhi, 2004.

Basu, S.K. and S.K. Bhattacharya, *Land Reforms in West Bengal: A Study in Implementation*, Oxford Book Company, Calcutta, 1963.

Basu, Tara Krishna, *The Bengal Peasant from Time to Time*, Calcutta, 1962.

Benda, Harry J., 'Non-Western Intelligentsia as Political Elites', in J.H. Kaustsky (ed.), *Political Change in Underdeveloped Countries*, New York, 1962.

Beloff, Max, 'Intellectuals' in Adam Kuper and Jessica Kuper (ed.), *The Social Science Encyclopaedia*, 1985.

Bhaduri, Amit, 'A Study of Agricultural Backwardness under Semi-feudalism', *The Economic Journal*, March 1973.

———, 'The Evolution of Land Relations in Eastern India Under British Rule', *Indian Economic and Social History Review*, January-March 1976.

Bharatkosh (relevant vols.), Bangiya Sahitya Parishad (Bengali), Calcutta, 1970.

Bharucha, Rustom, *Another Asia: Rabindranath Tagore & Okakura Tenshin*, Routledge, Delhi, 2006.

Bhattacharya, Bhabani, *Socio-Political Currents in Bengal: A Nineteenth Century Perspective*, Vikash Publishing House, Sahibabad, 1980.

Bhattacharya, Pasupati, *Antaranga Rabindranath* (Bengali), Mitra & Ghosh, Calcutta, 1994.

Bhattacharya, Pradyumna, 'Vidyasagar Ebong Besarkari Samaj' (Bengali), *Yogasutra*, July-September 1993.

Bhattacharya, Sabyasachi, 'Notes on the Role of the Intelligentsia in Colonial Society: India from Mid-nineteenth Century', *Studies in History*, vol. I, no. 1, January-June 1979.

——— (ed.), *The Mahatma and the Poet: Letters and Debates between Gandhi and Tagore 1915-1941*, National Book Trust, Delhi, 1997.

Bhattacharya, Shibaprasanna, *Vidyasagar Prabandha* (Bengali), S. Bhattacharya, Calcutta, 1305 B.S.

Bhattacharya, Subhas, 'The Indigo Revolt of Bengal', *Social Scientist*, vol. 5, no. 12, July 1977.

Bishi, Pramathanath (ed.), *Bhudeb Rachana Sambhar* (Bengali), Sahitya Samsad, Calcutta, 1364 B.S.

Blyn, George, *Agricultural Trends in India, 1891-1947: Output, Availability and Productivity*, University of Pennsylvania Press, Pennsylvania, 1966.

Bole, P.N. and H.W.B. Moreno, *A Hundred Years of the Bengali Press*, Calcutta, 1920.

Bose, Sugato, *Agrarian Bengal: Economy, Social Structure and Politics, 1919-1947*, Cambridge University Press, Cambridge, 1986.

———, *The New Cambridge History of India: Peasant Labour and Colonial Capital: Rural Bengal Since 1770*, Cambridge University Press, Cambridge, 1993.

Bottomore, T.B., *Elites and Society*, Penguin Books, London, 1966.

Bronowski, Jacob, *The Western Intellectual Tradition: From Leonardo to Hegel*, Hutchinson & Co., New York, 1960.

Broomfield, J.H., 'The Regional Elites: A Theory of Modern Indian History', *IESHR*, vol. II, no. 3, September 1966.

———, *Elite Conflict in a Plural Society: Twentieth Century Bengal*, University of California Press, Berkeley, 1968.

Buckland, C.E., *Bengal under the Lieutenant Governors*, vols. I and II, Calcutta, 1901.

Bullock, Alan and Oliver Stallybrass (eds.), *The Fontana Dictionary of Modern Thought*, Fontana Press, London, 1989.

Chakraborty, Bidyut, *Social and Political Thought of Mahatma Gandhi*, Routledge, Oxford, 2006.

Chakraborty, Dipesh, *Provincializing Europe: Postcolonial Thought and Historical Difference*, Princeton University Press, Princeton, 2000.

Chakravarty, Satyadas, *Sriniketaner Govar Katha* (Bengali), Sahitya Samaj, Sriniketan, 1985.

Chandra, Bholanath, *Raja Digambar Mitra: His Life and Work*, Calcutta, 1906.

Chandra, Bipan, *The Rise and Growth of Economic Nationalism in India*, People's Publishing House, Delhi, 1966.

Chatterjee, Asok Kumar, 'Influence of Tenancy Legislation in Bengal's Economy (1858-1939)', unpublished Ph.D. thesis, Rabindra Bharati University, 1986.

Chatterjee, Partha, 'Agrarian Structure in Pre-Partition Bengal', in Asok Sen, Partha Chatterjee and Sugata Mukherjee, *Three Studies on the Agrarian Structure in Bengal 1850-1947*, Oxford University Press, Delhi, 1984.

———, *Bengal 1920-147*, vol. 1, *The Land Question*, K.P. Bagchi & Co., Calcutta, 1984.

———, *Nationalist Thought and the Colonial World: A Derivative Discourse*, Zed Books, London, 1986.

———, *The Nation and its Fragments: Colonial and Post-Colonial Histories* Princeton University Press, Princeton, 1993.

——— (ed.), *Texts of Power: Emerging Disciplines in Colonial Bengal*, Samya, Calcutta, 1996.

———, 'Rabindrik Nation ki?', Puja edition of *Baromas*, 2003.

———, 'Rabindrik Nation Prasange Aro Du-char Katha', *Baromas*, 2004.

Chaudhuri, Abul Ahsan, *Meer Masarraf Hussain: Sahitya Karma O Samaj Chinta* (Bengali), Dacca, 1996.

Chaudhuri, Binay Bhusan, *Growth of Commercial Agriculture in Bengal, 1757-1900*, Indian Studies Past and Present, Calcutta, 1964.

———, 'Agrarian Economy and Agrarian Relations in Bengal 1859-1885', in N.K. Sinha (ed.), *The History of Bengal (1757-1905)*, University of Calcutta, Calcutta, 1968.

———, 'Agricultural Production in Bengal and Bihar: Co-existence of Decline and Growth', *Bengal Past and Present*, vol. 88, no. 2 (166), July-December 1969.

———, 'Growth of Commercial Agriculture in Bengal, 1859-1885', *IESHR*, vol. 7, no. 1, March 1970.

———, 'Peasant Movements in Bengal, 1850-1900', *Nineteenth Century Studies*, July 1973.

———, 'The Story of a Peasant Revolt in a Bengal District', *Bengal Past and Present*, July-December 1973.

———, 'The Agrarian Question in Bengal and the Government, 1850-1900', *The Calcutta Historical Journal*, vol. I, no. 1, July 1976.

———, 'Agricultural Growth in Bengal and Bihar, 1770-1900', *Bengal Past and Present*, January-July 1976.

———, 'Movement of Rent in Eastern India, 1773-1939', *Indian Historical Review*, July 1977.

———, 'Agrarian Relations, Eastern India', in Dharma Kumar (ed.), *The Cambridge Economic History of India*, vol. 2, *c.1757-c.1970*, Orient Longman in association with Cambridge University Press, Delhi, 1984.

———, 'Peasantry as a Category in Indian History', in Bharati Ray and David Taylor (eds.), *Politics and Identity in South Asia*, K.P. Bagchi & Co., Kolkata, 2001.

———, *History of Science, Philosophy and Culture in Indian Civilization*, vol. VIII, part 2: *Peasant History of Late Pre-Colonial and Colonial India:* Pearson Longman, Delhi, 2008.

Chaudhuri, Rosinka, *Gentlemen Poets in Colonial Bengal: Emergent Nationalism and the Orientalist Project*, Seagull Books, Calcutta, 2002.

Choudhury, Indira, *The Frail Hero and Virile History: Gender and the Politics of Culture in Colonial Bengal*, Oxford University Press, Delhi, 1998.

Chuckarberty, T.D., *Thoughts on Popular Education in Bengal*, Calcutta, 1870.

Collingham, E.M., *Imperial Bodies: The Physical Experience of the Raj, c.1800-1947*, Polity Press, London, 2001.

Cooper, Adrienne, *Sharecropping and Sharecroppers' Struggles in Bengal, 1930-1950*, K.P. Bagchi & Co., Calcutta, 1988.

Dalton, Denis, *Mahatma Gandhi: Non-Violent Power in Action*, Columbia University Press, New York, 1993.

Das, Harihar, *Life and Letters of Toru Dutt*, Oxford University Press, London, 1921.

Das, Sisir Kumar, *The Artist in Chains: The Life of Bankim Chandra Chatterjee*, New Statesman Publishing Co., Delhi, 1984.

Das Gupta, Ajit, *A History of Indian Economic Thought*, Rouledge, London, 1993.

Das Gupta, Subrata, *Jagadish Chandra Bose and the Indian Response to Western Science*, Oxford University Press, Delhi, 1999.

Das Gupta, Sugata, *A Poet and a Plan*, Thacker Spink & Co., Calcutta, 1933.

Das Gupta, Tapati, *Social Thought of Rabindranath Tagore*, Abhinav Publications, Delhi, 1993.

Das Gupta, Uma, 'Rabindranath Tagore on Rural Reconstruction: The Sriniketan Programme, 1921-41', *Indian Historical Review*, vol. IV, no. 1, January 1978.

———, *Rabindranath Tagore: A Biography*, Oxford University Press, Delhi, 2004.

Davidoff, Leonore, *Worlds between Historical Perspectives on Gender and Class*, Routledge, New York, 1995.

———, 'Landscape with Figures: Home and Community in English Society' (with Jeanne L'Esperance and Howard Newby), in *Worlds between Historical Perspectives on Gender and Class*, Routledge, New York, 1995.

De, Amalendu, 'Indigo Plantation: A Source of Oppression', *Bengal Past and Present*, July-December 1963.

———, *Pakistan Prastab O Fazlul Huq* (Bengali), Calcutta, 1972.

———, *Bangali Buddhijibi O Bichchinatabad* (Bengali), Ratna Prakashan, Calcutta, 1381 B.S.

———, *Chirasthayi Bandobasta O Bangali Buddhijibi* (Bengali), Ratna Prakashan, Calcutta, 1981.

De, Barun, 'A Biographical Perspective on the Political and Economic Ideas of Rammohun Roy', in V.C. Joshi (ed.), *Rammohun Roy and the Process of Modernization in India*, Vikas, Delhi, 1975.

———, 'The Colonial Context of the Bengal Renaissance', in C.H. Philips and Mary Doreen Wainright (ed.), *Indian Society and the Beginning of Modernization c. 1830-1850*, Macmillan, London, 1976.

———, 'Historiographical Critique of Renaissance Analogues for Nineteenth Century India', in Barun De (ed.), *Perspectives in the Social Sciences I: Historical Dimensions*, Oxford University Press, Calcutta, 1977.

De, Jatindranath, 'The History of the Krishak Praja Party of Bengal, 1929-47: A Study of Changes in Class and Inter-Community Relations in the

Agrarian Sector of Bengal', unpublished Ph.D. thesis, University of Delhi, 1977.

De, S.K., *Bengali Literature in the Nineteenth Century (1757-1857)*, revised edn., Firma K.L.M., Calcutta, 1962.

Deb, Chitra, *Antahpurer Atmakatha* (Bengali), Ananda Publishers, Calcutta, 1984.

Deb, Chittaranjan, *Sriniketan Parichay* (Bengali), Bichitra, Calcutta, 1368 B.S.

Desai, A.R., *Social Background of Indian Nationalism*, Popular Prakashan, Bombay, 1959.

Desai, A.R., ed., *Peasant Struggles in India*, Oxford University Press, Delhi, 1979.

Devi, Maitrayee, *Mangpute Rabindranath* (Bengali), Calcutta, 1943.

Dimock, E.C. (ed.), *Bengal: Literature and History*, Asian Studies Centre, Michigan, 1967.

Dube, Saurabh, 'Peasant Insurgency and Peasant Consciousness', *Economic and Political Weekly*, XX, II, 16 March 1985.

Duncan, Ronald (ed.), *The Writings of Gandhi*, Rupa & Co., Delhi, 1990.

Duff, Alexander, *New Era of the English Language and English Literature in India*, Edinburgh, 1837.

Dutt, Bhabatosh, *The Evolution of Economic Thinking in India*, Federation Hall Society, Calcutta, 1962.

———, *Arthanitir Pathe* (Bengali), Calcutta, 1977.

Dutta, Krishna and Andrew Robinson, *Rabindranath Tagore: The Myriad-minded Man*, Bloomsbury, London, 1995.

Edwardes, Michael, *The Sahibs and the Lotus: The British in India*, Constable, London, 1988.

Elmhirst, L.K., *Poet and Plowman*, Visva-Bharati, Calcutta, 1975.

Engels, Dagmar, *Beyond Purdah? Women in Bengal, 1890-1930*, Oxford University Press, Delhi, 1999.

Engels, Dagmar and Shula Marks, *Contesting Colonial Hegemony: State and Society in Africa and India*, British Academic Press, London, 1994.

Fanon, Frantz, *The Wretched of the Earth*, Grove Press, London, 2004.

Faruqui, Zia-ul-Hasan, *The Deobund School and the Demand for Pakistan*, Asia Publishing House, Bombay, 1963.

Forbes, Geraldine, *The New Cambridge History of India: IV. 2: Women in Modern India*, Cambridge University Press, Cambridge, 1998.

Gallagher, Jack, Gordon Johnson and Anil Seal (ed.), *Locality Province and Nation: Essays on Indian Politics 1870-1940*, Cambridge University Press, Cambridge, 1991.

Gangopadhyay, Goursundur, *Atpoure Rabindranath* (Bengali), Orient Book Co., Calcutta, 1968.

Ganguli, Birendranath, *Trends of Population and Agriculture in the Gangaetic*

Valley: A Study in Agricultural Economics, Metheun & Co. Ltd., London, 1938.

Ganguli, B.N., *Indian Economic Thought: Nineteenth Century Perspectives*, MacGraw Hill, Delhi, 1977.

Ghosh, Anindita, *Popular Publishing and the Politics of Language and Culture in a Colonial Society, 1778-1905*, Oxford University Press, New Delhi, 2006.

Ghosh, Benoy, *Banglar Samajik Itihaser Dhara* (Bengali), Ayan Publishing, Calcutta, 1968.

———, *Banglar Biddyotsamaj* (Bengali), Ayan Publishing, Calcutta, 1973.

———, *Vidyasagar O Bangali Samaj* (Bengali), Orient Longman, Calcutta, 1973.

Ghosh, J.C., *Literature in Bengal*, Oxford University Press, Oxford, 1942.

Ghosh, Kalimohan, 'Sriniketan Pallisangathan', *Bhandar*, May 1975 (rpt.).

Ghosh, Sisir, *Indian Sketches*, Patrika Office, Calcutta, 1898

Ghosh, Sujata, 'The British Indian Association (1851-1900)', *Bengal Past and Present*, vol. LXXVII, part II, 1958.

Ghosh, Yatindrakumar (comp.), Bengal Provincial Conference: Mymensingh session (1905); Berhampore session (1907); Dacca session (1913); Calcutta session (1917), Firma K.L.M., Calcutta, 1970-4.

Goswami, Onkar, 'Calcutta's Economy 1918-1970: The Fall from Grace', in Sukanta Chaudhuri (ed.), *Calcutta: The Living City*, vol. II, Oxford University Press, Calcutta, 1990.

Gramsci, Antonio, *Selections from the Prison Notebook* (tr. Quentin Hoare and Geoffrey Nowell Smith), International Publishers, New York, 1971.

Great Soviet Encyclopedia, vol. 10, New York, 1976.

Guha, Aurobindo (ed.), *Unpublished Letters of Vidyasagar*, Calcutta, 1971.

Guha, Ranajit, *A Rule of Property for Bengal: An Essay on the Idea of Permanent Settlement*, Mouton & Co., Paris, 1963.

———, 'Neel Darpan: The Image of the Peasant Revolt in a Liberal Mirror', *Journal of Peasant Studies*, 2, October 1974.

———, *Elementary Aspects of Peasant Insurgency in Colonial India*, Oxford University Press, Delhi, 1994.

———, 'Prose of Counter-Insurgency', in Ranajit Guha (ed.), *Subaltern Studies*, vol. II, Oxford University Press, Delhi, 1983.

———, 'Discipline and Mobilize', in Partha Chatterjee and Gyanendra Pandey (eds.), *Subaltern Studies*, vol. VII, Oxford University Press, Delhi, 1993.

———, 'Dominance without Hegemony and its Historiography', in Ranajit Guha (ed.), *Subaltern Studies*, vol. VI, Oxford University Press, Delhi, 1983.

Gupta, Atul Chandra, *Jamir Malik* (Bengali), Visva-Bharati, Calcutta, 1351 B.S.

Gupta, Bipinbihari, *Puratan Prasanga* (Bengali), Vidyabharati Sanskaram, Calcutta, 1996.

Gupta, Dhruba, 'Keno Meer Masarraf Hussain', *Anustoop* (Bengali), 29th year, no. 1, 1994.

Gupta, J.N. (ed.), *Life and Work of Romesh Chandra Dutt*, E.P. Dutton & Co., New York, 1911.

Hamid, Abdul, *Muslim Separatism in India: A Brief Survey, 1858-1947*, Punjab, 1967.

Hardy, Peter, *The Muslims of British India*, Cambridge University Press, London, 1972.

Hasan, Mushirul and Asim Roy (ed.), *Living Together Separately: Cultural India in History and Politics*, Oxford University Press, Delhi, 2005.

Hobsbawm, Eric J., *Industry and Empire* (The Pelican Economic History of Britain, vol. 3), Penguin Books, London, 1969.

Islam, M. Mufakharul, *Bengal Agriculture 1920-1947: A Quantitative Study*, Sultan Chand & Co., Delhi, 1978.

Islam, Sirajul, *The Permanent Settlement in Bengal: A Study of its Operation 1790-1819*, Bangla Academy, Dhaka, 1979.

———, 'Bengal Agrarian Society: Continuity and Change under the Colonial Rule', in Nisith Ranjan Ray et al. (eds.), *Agrarian Bengal under the Raj*, Calcutta, 1986.

———, *Bengal Land Tenure: The Origin and Growth of Intermediate Interests in the 19th Century*, K.P. Bagchi & Co., Calcutta, 1988.

Jaritz, Gerhard, 'The Material Culture of the Peasantry in the Late Middle Ages: "Image" and "Reality", in Del Sweeney (ed.), *Agriculture in the Middle Ages: Technology, Practice and Representations*, University of Pennsylvania, Philadelphia, 1995.

Johnson, Richard L. (ed.), *Gandhi's Experiments with Truth: Essential Writings by and about Mahatma Gandhi*, Oxford University Press, Oxford, 2006.

Joshi, V.C. (ed), *Rammohun Roy and the Process of Modernization in India*, Vikas, Delhi, 1975.

Kaviraj, Narahari, *Swadhinatar Sangrame Bangla* (Bengali), Calcutta, 1957.

———, 'Peasant Question' in Jagannath Chakravarti (ed.), *Studies in Bengal Renaissance*, National Council of Educations, Bengal, Calcutta, 1977.

———, *Wahabi and Farazi Rebels of Bengal*, People's Publishing House, Delhi, 1982.

Kaviraj, Sudipta, *The Unhappy Consciousness: Bankim Chandra Chattopadhyay and the Formation of National Discourse in India*, Oxford University Press, Delhi, 1995.

Kawai, Akinobu, *Landlords and Imperial Rule: Change in Bengal Agrarian Society, c. 1885-1940*, 2 vols., Institute for the Study of Languages and Cultures of Asia and Africa, Tokyo University of Foreign Studies, Tokyo, 1986-87.

Klein, Ira, 'Malaria and Mortality in Bengal, 1840-1921', *IESHR*, vol. 9, no. 1, 1972.

Kling, Blair, *The Blue Mutiny: Indigo Disturbances in Bengal 1859-1862*, University of Pennsylvania Press, Philadelphia, 1967.

Lal, P.C., *Reconstruction and Education in Rural India*, George Allen & Unwin Ltd., London, 1932.

Lipset, Seymour Martin, 'Intellectuals', in E.F. Borgatta (ed.), *Encyclopaedia of Sociology*, vol. II, Macmillan, New York, 1992.

Mannan, Kazi Abdul, *Unish Sataker Sahitya Patra O Muslim-Manash* (Bengali), Bangla Academy, Dacca, 1959.

———, *Adhunik Bangla-Sahitye Muslim Sadhana* (Bengali), vol. 1, Rajshahi University Press, Rashahi, 1961.

Mannheim, Karl, *Ideology and Utopia: An Introduction to the Sociology of Knowledge*, Harvest Book, London, 1961.

Majumdar, B.B., *History of Indian Social and Political Ideas: From Rammohan to Dayananda*, Bookland Pvt. Ltd., Calcutta, 1967.

———, *Indian Political Associations and Reform of Legislature 1818-1917*, Calcutta, 1965.

Majumdar, J.K. (ed.), *Raja Rammohun Roy and Progressive Movements in India: A Selection of Records (1775-1845)*, Art Press, Calcutta, 1941.

McGuire, John, *The Making of a Colonial Mind: A Quantitative Study of the Bhadralok in Calcutta, 1857-1885*, Australian National University Press, Canberra, 1983.

Michels, Roberto, 'Intellectuals', in Edwin R.A. Seligman (ed.), *Encyclopaedia of the Social Sciences*, vol. VII, New York, 1959.

Misra, B.B., *The Indian Middle Classes: Their Growth in Modern Times*, Oxford University Press, Delhi, 1978.

Mitra, Indra, *Karunasagar Vidyasagar* (Bengali), Ananda Publishers, Calcutta, 1992.

Mitra, Subal Chandra, *Iswar Chandra Vidyasagar* (*Story of his Life and Work*), Calcutta, first published in 1902, rpt., Ashish Publishing House, Delhi, 1975.

Moore, Barrington, Jr., *Social Origins of Dictatorship and Democracy: Lord and Peasant in the Making of the Modern World*, Penguin Press, London, 1967.

Mukherjee, Mridula, 'Peasant Resistance and Peasant Consciousness in Colonial India: "Subalterns" and Beyond', *Economic and Political Weekly*, 23 (41-2), 8 and 15 October 1988.

———, 'Peasant Movements and Nationalism in the 1920s', in Bipan

Chandra, et al., *India's Struggle for Independence*, Penguin Books India, Delhi, 1989.

———, *Peasants in India's Non-Violent Revolution, Practice and Theory*, Sage, Delhi, 2004.

Mukherjee, Mukul, 'Impact of Modernization on Women's Occupation: A Case Study of the Rice-Husking Industry of Bengal', *IESHR*, vol. 20, no. 1, 1983.

Mukherjee, Nilmani, *A Bengal Zamindar: Joykrishna Mukherjee of Uttarpara and His Times*, Firma KLM, Calcutta, 1975.

Mukherjee, Radhakamal, 'The Humanistic Meaning of Natural Science', in a committee of Experts appointed by Dr. A.V. Baliga Memorial Committee (ed.), *Science and Society: Essays in Honour of Dr. A.V. Baliga*, Bombay, 1972.

Mukherjee, Radharaman, *Occupancy Rights: Its History and Incidents*, University of Calcutta, Calcutta, 1919.

Mukherjee, Sisir, 'Rabindranather Sriniketan: Pragatishil Krishibhabnar Adipeeth', *Udichi*, Pous, 1388 B.S.

Mukherjee, S.N., 'Class, Caste and Politics in Calcutta, 1815-1838', in Edmund Leach and S.N. Mukherjee (eds.), *Elites in South Asia*, Cambridge University Press Cambridge, 1970.

———, 'Bhadralok in Bengali Language and Literature', *Bengal Past and Present*, vol. XCV, part II, July-December 1976.

Mullick, Sukumar, *Rabindranather Pallichinta* (Bengali), Nabapatra Prakashan, Calcutta, 1989.

Nag, Kalidas, and Burman, Debajyoti (eds.), *English Works of Rammohun Roy*, part IV, Sadharan Brahmo Samaj, Calcutta, 1947.

Nakazato, Nariaki, 'Superior Peasants of Central Bengal and their Land Management in the Late Nineteenth Century', *Journal of the Japanese Association for South Asian Studies*, no. 2, 1991.

———, *Agrarian System in Eastern Bengal, c. 1870-1910*, K.P. Bagchi & Co., Calcutta, 1994.

Nandy, Ashis, *The Intimate Enemy: The Loss and Recovery of Self under Colonialism*, Oxford University Press, Delhi, 1983.

———, *The Illegitimacy of Nationalism: Rabindranath Tagore and the Politics of Self*, Oxford University Press, Delhi, 1998.

Natrajan, L., *Peasant Uprisings in India 1850-1900*, People's Publishing House Ltd., Bombay, 1953.

New Standard Encyclopaedia, vol. 8, Standard Educational Corporation, Chicago, 1948.

Palit, Chittabrata, *Tensions in Bengal Rural Society: Landlords, Planters and Colonial Rule (1830-60)*, Progresive Publishers, Calcutta, 1975.

———, *New Viewpoints on Nineteenth Century Bengal*, Calcutta, 1980.

Panikkar, K.N., ed., *National and Left Movements in India*, Vikas Publishing House, Delhi, 1980.

Parel, Anthony J. (ed), *Cambridge Texts in Modern Politics: M.K. Gandhi, Hind Swaraj and Other Writings*, Cambridge University Press, Delhi, 1997.

Pati, Biswamoy, *Issues in Modern Indian History for Sumit Sarkar*, Mumbai, 2000.

———, *Situating Social History, Orissa (1800-1997)*, Orient Longman, New Delhi, 2001.

Paul, Prasanta Kumar, *Rabijibani* (Bengali), vols. I-IX, Ananda Publishers, Calcutta, 1982-2003.

Pearson, William, 'Santiniketan', *Visva-Bharati Patrika*, Baisakh-Asar 1371 B.S.

Poddar, Aurobindo, *Bankim Manas* (Bengali), Calcutta, 1960.

———, *Renaissance in Bengal: Search for Identity*, Simla Institute of Advanced Study, Simla, 1977.

Pouchepadass, J., 'Peasant Classes in Twentieth Century Agrarian Movements in India', in E.J. Hobsbawm et al. (eds.), *Peasants in History: Essays in Honour of Daniel Thorner*, Oxford University Press, Calcutta, 1980.

Ray, Rajat Kanta, 'The Crisis of Bengal Agriculture: Dynamics of Immobility', *IESHR*, vol. 10, no. 3, 1973.

———, *Social Conflict and Political Unrest in Bengal, 1875-1927*, Oxford University Press, Delhi, 1984.

———, 'The Retreat of the Jotedars', *IESHR*, vol. 25, no. 2, 1988, pp. 235-47.

Ray, Rajat and Ratnalekha, 'The Dynamics of Continuity in Rural Bengal under the British Imperium', *IESHR*, vol. 10, no. 2, 1973.

———, 'Zamindars and Jotedars: A Study of Rural Politics in Bengal', *Modern Asian Studies*, vol. 9, no. 1, 1975.

Ray, Ratnalekha, *Change in Bengal Agrarian Society 1750-1850*, Manohar, Delhi, 1979.

Ray, 'Suprakash', *Bharater Krishak Bidroha O Ganatantrik Samgram* (Bengali), Naya Prakash, Calcutta, 1970.

Roy Choudhury, Debiprasanna, 'Amader Nimnasreni O Durbhiksha', *Nabya Bharat*, Chaitra 1291 B.S.

Ray Chaudhuri, Tapan, *Europe Reconsidered: Perceptions of the West in Nineteenth Century Bengal*, Oxford University Press, Delhi, 1989.

Robb, Peter, *Meanings of Agriculture: Essays in South Asian History and Economics*, Oxford University Press, Delhi, 1996.

Rothermund, Dietmar, *Government, Landlord and Peasant in India: Agrarian Relations under British Rule, 1865-1935*, Franz Steiner Verlag, Wiesbaden, 1978.

Sanyal, Hites, 'Congress Movements in the Villages of Eastern Midnapore, 1921-1931', *Asie du Sud Traditions et Changements, Colloques Internationaux du Centre National de la Recherche Scientifique*, 582, Paris.

Sarkar, Susobhan Chandra (ed.), *Rammohun Roy on Indian Economy*, People's Publishing House, Calcutta, 1965.
———, *Bengal Renaissance and other Essays*, People's Publishing House, Delhi, 1973.
Sarkar, Sumit, *The Swadeshi Movement in Bengal, 1903-1908*, People's Publishing House, Delhi, 1973.
———, 'The Complexities of Young Bengal', *Nineteenth Century Studies*, no. 4, 1973.
———, 'Rammohun Roy and the Break with the Past', in V.C. Joshi (ed.), *Rammohun Roy and the Process of Modernization*, Vikas, Delhi, 1975.
———, *Modern India, 1885-1947*, Macmillan, Madras, 1983.
———, *Popular Movements and Middle Class Leadership in Late Colonial India: Perspectives and Problems of a History from Below*, Calcutta, 1983.
———, *Writing Social History*, Oxford University Press, Delhi, 1997.
———, *Beyond Nationalist Frames: Relocating Postmodernism, Hindutva, History*, Permanent Black, Delhi, 2002.
Schendel, William van, *Peasant Mobility: The Odds of Life in Rural Bangladesh*, Manohar, Delhi, 1982.
Scott, James C., *The Moral Economy of the Peasant: Rebellion and Subsistence in Southeast Asia*, Yale University Press, New Haven, 1976.
———, *Weapons of the Weak: Everyday Forms of Peasant Resistance*, Oxford University Press, Delhi, 1990.
———, *Domination and the Arts of Resistance: Hidden Transcripts*, Yale University Press, New Haven, 1990.
Seal, Anil, *The Emergence of Indian Nationalism: Competition and Collaboration in the Late 19th Century*, Cambridge University Press, Cambridge, 1971.
Sen, Asok, *Iswar Chandra Vidyasagar and His Elusive Milestones*, Riddhi, Calcutta, 1977.
———, 'Agrarian Structure and Tenancy Laws in Bengal, 1850-1885', in Asok Sen, Partha Chatterjee and Saugata Mukherjee, *Three Studies on the Agrarian Structure in Bengal: 1850-1947*, Oxford University Press Calcutta, 1982.
Sen, Asoka Kumar, *The Popular Uprising and the Intelligentsia: Bengal Between 1855-1873*, Firma KLM, Calcutta, 1992.
Sen, Bhowani, *Evolution of Agrarian Relations in India*, People's Publishing House, Delhi, 1962.
Sen, Sachin, *Banglar Raiyat O Zamindar* (Bengali), Visva-Bharati, Calcutta, 1944.
Sen, Sudhir, *Rabindranath Tagore on Rural Reconstruction*, Visva-Bharati, Calcutta, 1978.
Sen, Sunil, *Agrarian Struggle in Bengal: 1946-47*, People's Publishing House, Delhi, 1972.
———, *Peasant Movements in India*, Calcutta, 1982.

Sengupta, Kalyan Kumar, 'The Agrarian League of Pabna, 1873', *IESHR*, vol. 7, no. 2, 1970.

———, 'Agrarian Disturbances in Nineteenth Century Bengal', *IESHR*, vol. 8, no. 2, June 1971.

———, 'Peasant Struggle in Pabna: It's Legalistic Character', *Nineteenth Century Studies*, no. 3, July 1973.

———, *Pabna Disturbances and the Politics of Rent, 1873-1885*, People's Publishing House, Delhi, 1974.

———, Violence in Rural Bengal, Zamindar and Peasants, 1859-1947, *Bengal Past and Present*, July-December, 1983.

Sengupta, Pramode, *Nil Bidroha O Bangali Samaj* (Bengali), National Book Agency Ltd., Calcutta, 1960.

Sharp, H. (comp. & ed.), *Selections from Educational Records*, part-I, *1781-1830*, Calcutta, 1920, reprinted for the National Archives of India, Delhi, 1965.

Sastri, Sibnath, *Ramtanu Lahiri O Tatkalin Banga Samaj* (Bengali), New Age Publishers, Calcutta, 1955.

———, *History of the Brahmo Samaj* (1st published in 1911), Sadharan Brahmo Samaj, Calcutta, 1974.

Shanin, Theodor (ed.), *Peasants and Peasant Societies*, Penguin Books, Middlesex, 1971.

Shils, E.A., 'Political Development in the New States', *Comparative Studies in Society and History*, vol. II, New York, 1960.

———, 'Intellectuals', in David L. Sills (ed.), *International Encyclopedia of the Social Sciences*, vol. VII, New York, 1972.

Singh-Roy, P.N., *The Chronicle of the British Indian Association*, Calcutta, 1965.

Sinha, N.K., *Economic History of Bengal*, vols. I and II, Firma KLM, Calcutta, 1962 and 1970.

Sinha, Pradip, *Nineteenth Century Bengal: Aspects of Social History*, Firma KLM, Calcutta, 1965.

Sinha, Sasadhar, *Social Thinking of Rabindranath Tagore*, Asia Publishing House, Calcutta, 1962.

Stokes, Eric, *The English Utilitarians and India*, Clarendon Press, Oxford, 1959.

———, *The Peasant and the Raj: Studies on Agrarian Society and Peasant Rebellion in Colonial India*, Cambridge University Press, Cambridge, 1978.

Sweeney, Del (ed.), *Agriculture in the Middle Ages: Technology, Practice and Representations*, University of Pennsylvania, Philadelphia, 1995.

Sykes, Marjorie, *Rabindranath Tagore*, Thacker Spink & Co., Calcutta, 1963.

Tagore, Kshitindranath, *Dwarkanath Thakurer Jibani* (Bengali), Calcutta, 1376 B.S.

Tagore, Pratima, 'Smritichitra', *Visva-Bharati Patrika*, Sravana-Aswin 1352 B.S.

Thompson, E.P., *Alien Homage: Edward Thompson and Rabindranath Tagore*, Oxford University Press, Delhi, 1993.

Thorner, Daniel and Alice Thorner, *Land and Labour in India*, Asia Publishing House, Bombay, 1962.

Tripathi, Amales, *Vidyasagar: The Traditional Modernizer*, Orient Longman, Bombay, 1974.

Umar, Badruddin, 'Ishwar Chandra Vidyasagar', in Ghulam Murshid (ed.), *Vidyasagar*, Chirayata Prakashan, Calcutta, 1971.

———, *Iswar Chandra Vidyasagar O Unish Sataker Bangali Samaj* (Bengali), Chirayata Prakashan, Calcutta, 1980.

———, *Chirasthayi Bandobasta Bangladesher Krishak* (Bengali), Chirayata, Calcutta, 1974.

Venturi, Franco, *Roots of Revolution*, Universal Library, New York, 1966.

Wolf, Eric R., *Peasant Wars of the Twentieth Century*, Faber and Faber, London, 1971.

Woodrow, H. (coll. & ed.), *Macaulay's Minutes on Education in India*, Calcutta, 1902.

Index